STECK-VAUGHN
TOP LINE Math

Ratios and Percents

Teacher's Guide

You can download free resources for your students at
www.HarcourtAchieve.com/AchievementZone.

Rigby • Saxon • Steck-Vaughn

www.HarcourtAchieve.com
1.800.531.5015

Acknowledgments

Editorial Director Ellen Northcutt
Supervising Editor Pamela Sears
Senior Editor Kathy Immel

Associate Design Director Joyce Spicer
Design Team Jim Cauthron
Joan Cunningham

Photo Researcher Stephanie Arsenault

Cover Art ©Janet Parke

ISBN 1-41-900379-8

© 2006 Harcourt Achieve Inc.
All rights reserved. No part of the material protected by this copyright may be reproduced or utilized in any form or by any means, in whole or in part, without permission in writing from the copyright owner. Requests for permission should be mailed to: Copyright Permissions, Harcourt Achieve, P.O. Box 27010, Austin, Texas 78755. Rigby and Steck-Vaughn are trademarks of Harcourt Achieve Inc. registered in the United States of America and/or other jurisdictions.

1 2 3 4 5 6 7 8 9 10 048 11 10 09 08 07 06 05 04

Contents

Top Line Math Components. iv

Using this Teacher's Guide. vii

UNIT 1
Ratio and Proportion — 2

Overview Lessons 1–3
Ratios and Rates 3
   Lesson 1 Ratios 4
   Lesson 2 Rates 5
   Lesson 3 Learning About Unit Rates 6

Overview Lessons 4–5
Proportions 7
   Lesson 4 Proportions 8
   Lesson 5 Using Cross Products 9

Overview Lessons 6–7
More on Proportions 10
   Lesson 6 Writing Proportions 11
   Lesson 7 Problem Solving Using Proportions 12

Unit 1 Review **13**

UNIT 2
Fractions, Decimals, and Percents — 14

Overview Lessons 8–10
Fractions and Percents 15
   Lesson 8 Understanding Percents 16
   Lesson 9 Converting Fractions to Percents 17
   Lesson 10 Converting Percents to Fractions 18

Overview Lessons 11–12
Decimals and Percents 19
   Lesson 11 Converting Decimals to Percents 20
   Lesson 12 Converting Percents to Decimals 21

Overview Lessons 13–15
Percents and Problem Solving I 22
   Lesson 13 Find the Part 23
   Lesson 14 Find the Whole 24
   Lesson 15 Find the Percent 25

Overview Lessons 16–17
Percents and Problem Solving II 26
   Lesson 16 What Is a Percent Equation? 27
   Lesson 17 Creating a Percent Equation from a Word Problem. 28

Unit 2 Review **29**

UNIT 3
Percents in Daily Life — 30

Overview Lessons 18–19
Application of Percents I 31
   Lesson 18 Percent Change 32
   Lesson 19 Markup and Discount 33

Overview Lessons 20–21
Application of Percents II 34
   Lesson 20 Simple Interest 35
   Lesson 21 Compound Interest 36

Unit 3 Review **37**

Answers and Explanations 38

Copy Masters 49

Pretest/Post Test Evaluation Chart 55

STECK-VAUGHN
TOP LINE MATH

Ensure Student Success with Steck-Vaughn *Top Line Math*.

- Reach students struggling with math content and reading level.
- Build basic foundational math skills.
- Diagnose specific math intervention needs.
- Provide differentiated instruction.

1. Place your students in the right book.

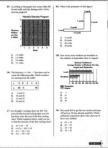

Give the Diagnostic to determine which of the ten *Top Line Math* books each student should use.

2. Diagnose.

Use the Pretest in each book to determine individual student's strengths and weaknesses.

Students receive exactly the instruction and practice they need. With *Top Line Math*, students will

- develop math competence.
- acquire basic math skills and concepts.
- learn problem-solving strategies.
- apply these skills and strategies to everyday life.
- gain confidence in their own ability to succeed at learning.

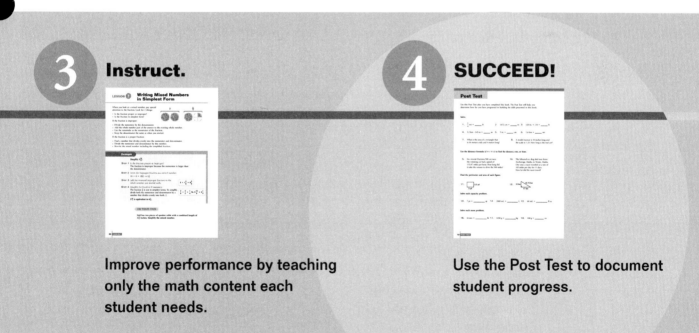

3 Instruct.

Improve performance by teaching only the math content each student needs.

4 SUCCEED!

Use the Post Test to document student progress.

COMPONENTS

Ten Student Books with Teacher's Guides make it possible for you to prescribe specific instruction for individual students based on their performance on the *Top Line Math* Diagnostic.

The Student Books

- Number Concepts
- Fractions
- Decimals
- Ratios and Percents
- Measurement
- Geometry
- Data, Tables, and Graphs
- Pre-Algebra
- Data Analysis and Probability
- Mathematical Reasoning

Each student book contains

- goal-setting activities that encourage students to take ownership of their learning
- engaging topics that link math to the real world
- step-by-step lessons that help students progress with confidence
- test-taking strategies that ensure students' success on key assessments
- tips to help students avoid common errors
- unit reviews that assesses mastery of specific skills

THE TEACHER'S GUIDE

The **Top Line Math** Teachers' Guide is your tool for delivering effective instruction. Get everything you need to improve student performance in one easy-to-use book.

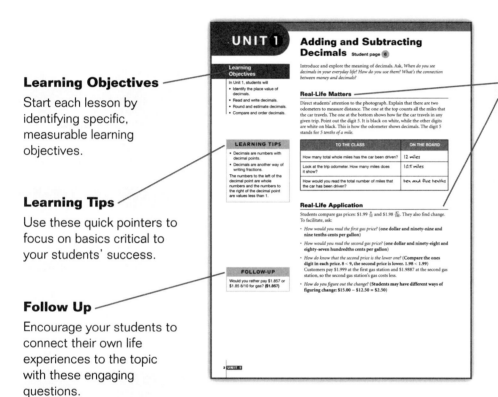

Learning Objectives
Start each lesson by identifying specific, measurable learning objectives.

Learning Tips
Use these quick pointers to focus on basics critical to your students' success.

Follow Up
Encourage your students to connect their own life experiences to the topic with these engaging questions.

Real-Life Matters and Real-Life Application
Tie learning to the real world. Use the chart as a springboard for presenting content.

Focus on specific skills grouped by key math concepts

Access Prior Knowledge
Motivate your students by using their prior knowledge to introduce key math concepts.

Remember the Basics
Review foundational skills before you introduce new content.

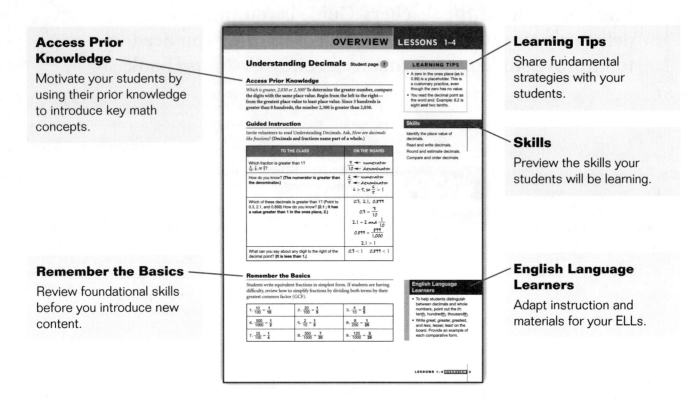

Learning Tips
Share fundamental strategies with your students.

Skills
Preview the skills your students will be learning.

English Language Learners
Adapt instruction and materials for your ELLs.

Learning Styles
Tailor instruction to meet the needs of the diverse students in your class.

Test-Taking Strategy
Prepare your students for success in important test-taking situations

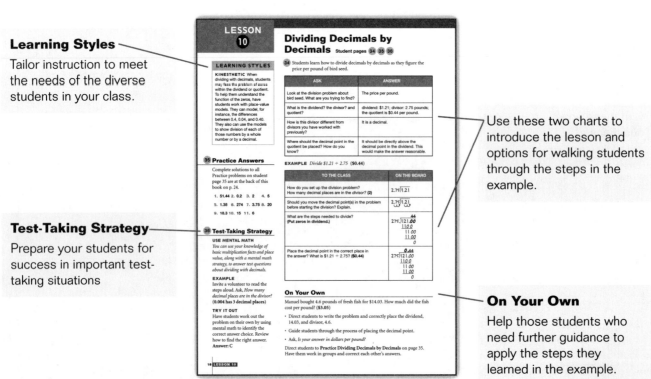

Use these two charts to introduce the lesson and options for walking students through the steps in the example.

On Your Own
Help those students who need further guidance to apply the steps they learned in the example.

When some of your students need additional instruction and others are ready to move on, *Top Line Math* provides activities that focus on all of your students' individual needs.

Review Test-Taking Strategies

Identify and explain test-taking strategies to ensure success on standardized tests.

Revisit Learning Objectives

Review the learning objectives on the Overview page so students can track their progress

Reteach

Use these activities for even more classroom practice or for homework

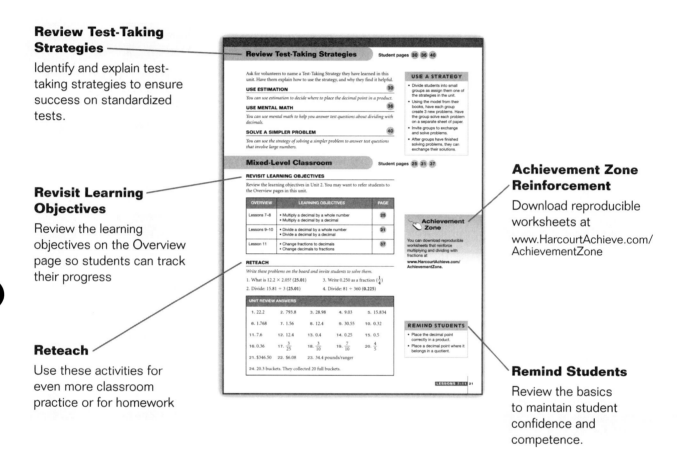

Achievement Zone Reinforcement

Download reproducible worksheets at www.HarcourtAchieve.com/AchievementZone

Remind Students

Review the basics to maintain student confidence and competence.

UNIT 1

Ratio and Proportion — Student page 6

Students explore ratio and proportion by considering prices for DVDs at two competing stores in a mall.

Learning Objectives

In Unit 1, students will
- Use ratios to compare two quantities.
- Reduce ratios to their simplest form.
- Learn that rates and unit rates are special kinds of ratios.
- Set up a proportion.
- Use cross products.
- Write proportions for everyday situations.
- Use equivalent ratios and proportional reasoning to solve problems.

Real-Life Matters

Invite volunteers to read the Real-Life Matters on page 6. Students use ratio and proportion to do some comparative pricing of DVD rentals.

ASK	ANSWER
What type of deal are the stores in the mall offering on DVD rentals?	Movie Time is offering 4 DVD rentals for $15. DVD World rents DVDs for $4.50 each.
How can you determine the rental fee for one DVD at Movie Time?	Divide $15 by 4. $15 ÷ 4 = $3.75 per DVD rental.
Which rental company offers the better deal?	Movie Time does, if you want to rent 4 DVDs at a time.
A **ratio** compares two numbers. You can write the Movie Time rental offer as a ratio of $\frac{15}{4}$. What is this ratio comparing?	The ratio compares the rental price to the number of DVDs rented.
A **proportion** means that two ratios are equal. Are these two ratios ($\frac{15}{4}$ and $\frac{3.75}{1}$) equal? Explain why or why not.	Yes, because you can think of equivalent fractions to see that these two ratios are equal: $\frac{15}{4} \div \frac{4}{4} = \frac{3.75}{1}$

LEARNING TIPS

- A fraction is another way of expressing a division problem. Example: $\frac{15}{4} = 15 \div 4 = \$3.75$
- Ratios can be written as fractions.

Real-Life Application

Let students share ideas on how they work with ratios and proportions in real-life situations by brainstorming answers to the questions presented.

- *Another DVD rental company charges $21 per month for unlimited DVD rentals. If you rented 7 DVDs per month, what is the rental cost per DVD?* ($\frac{21}{7} = \$3$ per DVD rental)

- *If you rented 14 DVDs per month, what would the cost per DVD rental be?* ($\frac{21}{14} = \$1.50$ per DVD rental)

- *What if you found another DVD rental plan for $3 per DVD and this was the lowest price you could find? You have a $21 allowance per month for DVD rentals. How many DVDs could you rent each month with this plan?* ($\frac{21}{3} = 7$; you could rent 7 DVDs per month with this plan.)

- *Which rental plan would you choose and why?* (**Invite students to explain their choices to the class.**)

FOLLOW-UP

Which is the better deal, renting 5 DVDs for $4, or 10 DVDs for $8? (**Neither. They have the same cost per DVD rental.** $\frac{5}{4} = \frac{10}{8} = \$1.25$ per rental)

OVERVIEW LESSONS 1–3

Ratios and Rates Student page 7

Access Prior Knowledge

What fraction of the class wears glasses? Write a fraction using the number of students wearing glasses and the total number of students. **Answers will vary.**

Guided Instruction

Direct students to the Overview instruction that displays a pie consisting of 6 slices; 2 slices are blueberry and the other 4 slices are apple.

TO THE CLASS	ON THE BOARD
How many slices are in the pie?	6 slices
How many of the pie slices are blueberry?	2 blueberry
What fraction of the pie is made up of blueberry slices?	$\frac{2}{6}$
Equivalent means two things are equal. What equivalent fraction can you write for $\frac{2}{6}$?	$\frac{2}{6} = \frac{1}{3}$
A ratio compares two quantities. What is the ratio of blueberry slices to all the slices?	Point to $\frac{2}{6}$.
What is this ratio in simplest form?	Point to $\frac{1}{3}$.
Ratios can be written in three different ways: 1 to 3, $\frac{1}{3}$, 1 : 3. How does each of these 3 different ways express the ratio of blueberry slices to all the pie slices? (1 to 3 means 1 blueberry slice for every 3 slices of pie; $\frac{1}{3}$ shows this same relationship in fraction form; 1 : 3 identifies each value in this relationship and separates it with a colon.)	Write the ratio forms on the board. 1 to 3, $\frac{1}{3}$, 1 : 3

Remember the Basics

Direct students to the chart at the bottom of the page. In this activity, students fill in the empty spaces on the chart with equivalent fractions. See *Top Line Math: Fractions* if students need additional help. *Could you write more equivalent fractions for each fraction shown? Explain.* (**Yes; there are an infinite number of equivalent fractions that could be written for each fraction shown on the chart.**)

1. $\frac{8}{10} = \frac{4}{5} = \frac{16}{20}$	2. $\frac{3}{7} = \frac{6}{14} = \frac{9}{21}$
3. $\frac{5}{12} = \frac{10}{24} = \frac{15}{36}$	4. $\frac{1}{4} = \frac{2}{8} = \frac{3}{12}$
5. $\frac{12}{18} = \frac{6}{9} = \frac{2}{3}$	6. $\frac{3}{8} = \frac{6}{16} = \frac{9}{24}$

Skills

Use ratios to compare two quantities.

Reduce ratios to their simplest form.

Learn that rates and unit rates are special kinds of ratios

KEY WORDS

ratio: compares two quantities

proportion: a statement that says two ratios are equal

equivalent: when two things are equal

rate: a ratio that compares two quantities measured in different units

unit rate: the rate for one unit of a quantity. The denominator for any unit rate is 1.

LEARNING TIPS

When expressing ratios,
- First write the ratio with words: 2 blueberry slices to 6 slices of pie.
- Next write the statement in part/whole format:
  $\frac{2 \text{ blueberry slices}}{6 \text{ slices of pie.}}$
- Finally show the ratio with numerical values: $\frac{2}{6}$.

English Language Learners

- Whenever possible, show a visual of the items being compared. For example, 6 DVD rentals for $12. Let students see all 12 dollars and the 6 DVDs so students can directly experience pairing dollars with DVDs to express the ratio: $\frac{1 \text{ DVD}}{\$2}$

LESSON 1

Ratios Student pages 8 9

8 Students learn how to read and write ratios in three different ways.

TO THE CLASS	ON THE BOARD
A ratio compares two like quantities or numbers. The ratio of keys on a piano is 5 black keys to 7 white keys.	5 black keys to 7 white keys
How do you write this as a ratio? You say each ratio the same way: 5 to 7	5 to 7, 5 : 7, $\frac{5}{7}$

EXAMPLE *This basketball team has 9 players. Look at the uniforms. Write a ratio comparing the number of uniforms that have odd numbers to the number of uniforms that have even numbers.* (**2 to 1, 2 : 1, and $\frac{2}{1}$**)

ASK	ANSWER
How many of the uniforms have odd numbers?	6 uniforms
How many of the uniforms have even numbers?	3 uniforms
Write a statement comparing odd-numbered and even-numbered uniforms.	6 odd uniforms to 3 even uniforms
What ratio can you write for this statement?	6 to 3, or 6 : 3, or $\frac{6}{3}$
Is this ratio in simplest form?	No
What is the Greatest Common Factor for 6 and 3?	3
What is the ratio in simplest form?	$\frac{6}{3} \div \frac{3}{3} = \frac{2}{1}$
What is the ratio of odd-numbered uniforms to even-numbered uniforms?	2 to 1, 2 : 1, and $\frac{2}{1}$

On Your Own

There are 24 students in the school musical. Ten of the students are seniors. What is the ratio of students who are seniors to students who are *not* seniors? (**5 : 7**) Guide students to carefully read the problem.

- Ask, *What are you asked to compare in this problem?* (**number of seniors to number of non-seniors**)

- Direct students to compare the number of seniors/non-seniors. (**There are 10 seniors compared to 14 non-seniors.**)

- Invite a volunteer to explain how he or she simplified the ratio. ($\frac{10}{14} = \frac{5}{7}$)

Direct students to **Practice Ratios** on page 9. Have them work in groups and correct each other's answers.

LEARNING STYLES

KINESTHETIC Have the class create groupings of students based on different characteristics and then make ratios for the comparisons the groupings show. Example: Group by number of students who are the oldest sibling to those who are not; the number of left-handed students to right-handed students, etc.

9 Practice Answers

Complete solutions to all Practice problems on student page 9 are at the back of this book on p. 38.

1. $\frac{2}{5}$, 2 to 5, 2 : 5
2. $\frac{2}{5}$, 2 to 5, 2 : 5
3. $\frac{6}{5}$, 6 to 5, 6 : 5
4. $\frac{6}{18} = \frac{1}{3}$, 1 to 3, 1 : 3
5. $\frac{5}{3}$, 5 to 3, 5 : 3
6. $\frac{5}{6}$, 5 to 6, 5 : 6
7. $\frac{3}{6} = \frac{1}{2}$, 1 to 2, 1 : 2
8. $\frac{6}{8} = \frac{3}{4}$, 3 to 4, 3 : 4
9. $\frac{7}{11}$, or 7 to 11, or 7 : 11
10. $\frac{25}{35} = \frac{5}{7}$, or 5 to 7, or 5 : 7
11. $\frac{8}{12} = \frac{2}{3}$
12. $\frac{7}{11}$ or 7 to 11 or 7 : 11

Rates Student pages 10 11

LESSON 2

10 Students learn how to read and write rates in simplest form.

TO THE CLASS	ON THE BOARD
A rate is a special type of ratio that compares two different units. Can you give me an example of a rate?	mph or miles per hour
How can you express the rate of 250 miles in 4 hours as a ratio?	$\frac{250 \text{ miles}}{4 \text{ hours}}$ or $\frac{250}{4}$
If someone can do 250 sit-ups in 30 minutes, how can you write this as a ratio?	$\frac{250 \text{ sit-ups}}{30 \text{ minutes}}$ or $\frac{250}{30}$

EXAMPLE *A sale sign reads: Big Sale on CDs Today! $24 for sets of 3. Write the sale price for CDs as a rate in simplest terms.* (**8 : 1**)

ASK	ANSWER
What are the two different quantities being compared?	Sale price for a number of CDs $24 for 3 CDs
How do you write this rate as a ratio?	$\frac{\$24}{3 \text{ CDs}}$
Is this ratio in simplest form?	No, find the GCF to write the ratio in simplest form. $\frac{\$24}{3 \text{ CDs}} \div \frac{3}{3} = \frac{\$8}{1 \text{ CD}}$
What is the rate for the sales price of the CDs?	The CDs are selling at a rate of $8 per CD or 8 : 1.

On Your Own

Marcus can flip a coin 140 times in 4 minutes. At what rate does he flip a coin?

- Guide students to identify the two quantities being compared. (**140 flips and 4 minutes**)
- Have students express these two quantities as a fraction. ($\frac{140 \text{ flips}}{4 \text{ minutes}}$)
- Remind students to find the simplest form of the answer. ($\frac{140 \text{ flips}}{4 \text{ minutes}} \div \frac{4}{4} = \frac{35 \text{ flips}}{1 \text{ minute}}$)

Direct students to **Practice Rates** on page 11. Have them work in groups and correct each other's answers.

LEARNING STYLES

VISUAL Have students write out the process of expressing a rate in simplest form by showing both the numerator and denominator divided by the same value. Tell students to circle the fraction that they use to simplify their rate (which has a value of one).

Practice Answers 11

Complete solutions to all Practice problems on student page 11 are at the back of this book on p. 38.

1. 5 miles in 3 hours
2. $10 for 3 books
3. 3 free throws in 8 attempts
4. 25 liters in 2 minutes
5. $8 per hour
6. 1 ticket for $16
7. 3 cups of flour for two eggs
8. $75 per tire
9. 2 laps per minute
10. $\frac{40 \text{ miles}}{1 \text{ hour}}$
11. $\frac{5 \text{ rolls}}{2 \text{ hours}}$
12. $\frac{3 \text{ laps}}{4 \text{ minutes}}$
13. $\frac{\$15}{2 \text{ days}}$
14. $\frac{15 \text{ squirrels}}{2 \text{ hours}}$
15. $\frac{1 \text{ gallon}}{17 \text{ miles}}$

LESSON 3

Learning About Unit Rates

Student pages 12 13 14

12 Students learn to find the unit rate.

TO THE CLASS	ON THE BOARD
What does a rate compare? (A rate compares two different quantities.)	$\frac{90 \text{ miles}}{2 \text{ hours}}$ $\frac{130 \text{ words}}{2 \text{ minutes}}$
What does a unit rate express? (A unit rate is the rate for one unit of a quantity.)	$\frac{90 \text{ miles}}{2 \text{ hours}} = 90 \div 2 = \frac{45 \text{ miles}}{1 \text{ hour}}$
(The denominator in a unit rate will always be 1.)	$\frac{130 \text{ words}}{2 \text{ minutes}} = 130 \div 2 = \frac{65 \text{ words}}{1 \text{ minute}}$
What is another example of a unit rate that you see in a store?	unit price $\frac{\$3.69}{\text{pound}}$

EXAMPLE *A box of cereal contains 9 servings and a total of 72 grams of fiber. How many grams of fiber are in one serving?* ($\frac{8 \text{ grams}}{1 \text{ serving}}$)

ASK	ANSWER
What quantities are you asked to compare?	grams of fiber to servings
How do you write this rate as a fraction?	$\frac{72 \text{ grams of fiber}}{9 \text{ servings}}$
How do you find the unit rate? (divide)	$\frac{72 \text{ grams of fiber}}{9 \text{ serving}} = 72 \div 9 = 8$
What is the rate of grams per serving?	$\frac{8 \text{ grams of fiber}}{1 \text{ serving}}$

On Your Own

Jamal drove 66 miles in $2\frac{1}{2}$ hours. How many *miles per hour* did he drive? Guide students to write the quantities as a fraction. ($\frac{66}{2\frac{1}{2}} = \frac{26.4 \text{ miles}}{1 \text{ hour}}$)

- Ask, *How do you find the unit rate?* (**Divide the numerator by the denominator.**)
- Invite a volunteer to solve the problem on the board and explain how he or she arrived at the solution.

Direct students to **Practice Learning About Unit Rates** on page 13. Have them work in groups and correct each other's answers.

LEARNING STYLES

AUDITORY Have students listen for the word *per*. Expressions such as *per hour* and *per minute* identify a unit rate. Have students keep a log of their observations for class discussion.

13 Practice Answers

Complete solutions to all Practice problems on student page 13 are at the back of this book on p. 39.

1. 15 miles per hour
2. $9 per hour
3. 50 words per minute
4. 44 miles per hour
5. 30¢ per can of juice
6. 6 grams of fat per serving
7. 300 meters per minute
8. 24 miles per gallon
9. 0.5 pages per minute
10. $\frac{\$0.40}{1 \text{ bottle}}$
11. $\frac{2 \text{ pages}}{1 \text{ minute}}$
12. $\frac{50 \text{ words}}{1 \text{ minute}}$
13. $\frac{\$1.50}{1 \text{ slice}}$
14. $\frac{\$87.50}{1 \text{ room}}$
15. $\frac{\$0.80}{1 \text{ pound}}$

14 Test-Taking Strategy

USE LOGICAL REASONING
You can create a Venn diagram to compare quantities.

EXAMPLE
Ask for volunteers to read aloud the example and steps.

TRY IT OUT
Have students work out the problem on their own. Review how to find the right answer.
Answer: D

6 LESSON 3

OVERVIEW | LESSONS 4–5

Proportions Student page 15

Access Prior Knowledge

If 4 bagels cost $3, what would you pay for 8 bagels? **You can find the cost of one bagel and multiply that by 8, or multiply $3 × 2 since you are buying twice as many bagels, or you can set up a proportion. You would pay $6 for 8 bagels.**

Guided Instruction

Direct students to the Overview instruction and chart that lists breakfast foods and prices.

ASK	ANSWER
The chart shows that $6 buys 1 dozen bagels. How many is 1 dozen?	12
If you can buy 12 bagels for $6, how much will 24 bagels cost?	$12
You can write this as $\frac{\$6}{12 \text{ bagels}} = \frac{\$12}{24 \text{ bagels}}$. What can you say about these ratios?	Write on board: $\frac{\$6}{12 \text{ bagels}} = \frac{\$12}{24 \text{ bagels}}$ They are equal.
A statement that says two ratios are equal is called a proportion. The chart shows that 10 sweet rolls cost $4. What would 5 sweet rolls cost? How do you know?	$2. Since you are buying half as many sweet rolls, the price should be half of $4, which is $2.
Are these ratios equal? What proportion could you write for these equal ratios?	Yes. Write on board: $\frac{10 \text{ sweet rolls}}{\$4} = \frac{5 \text{ sweet rolls}}{\$2}$

Remember the Basics

Direct students to the chart at the bottom of the page. In this activity, students fill in the empty spaces on the chart to express ratios in simplest form. If students are having difficulty, guide them to express each ratio as a fraction, and review rules for finding equivalent fractions using the GCF. *How can you tell that a ratio is in simplest form?* **(The GCF of both terms is 1.)**

RATIO	RATIO IN SIMPLEST FORM	RATIO	RATIO IN SIMPLEST FORM
1. 4 : 6	2 : 3	2. 14 to 28	1 : 2
3. $\frac{3}{9}$	1 : 3	4. $\frac{12}{20}$	3 : 5
5. 5 to 20	1 : 4	6. 10 : 24	5 : 12

Skills

Set up proportions.
Compute cross products.

KEY WORDS

proportion: a statement that says two ratios are equal

equivalent: two things are equal

cross multiply: to multiply the numerator of one ratio by the denominator of the other ratio in a proportion

cross products: the answer to multiplying the numerator of one ratio by the denominator of the other ratio in a proportion

LEARNING TIPS

- When cross multiplying, draw an arrow from the numerator in one ratio to the denominator in the other ratio.
- Remember that multiplication can be shown in different ways, such as 12 × n or 12n.

English Language Learners

- To help students visualize the values in each ratio that forms a proportion, have them place a picture in each ratio. For example: $\frac{\$}{\text{(picture) bagel}} = \frac{\$}{\text{(picture) bagel}}$. Then let them place the value next to the picture in each ratio.

LESSON 4

Proportions Student pages 16 17

LEARNING STYLES

AUDITORY Have students cross multiply aloud so they can hear the product of each cross multiplication. The auditory response will reinforce whether ratios are alike or different.

16 Students learn how to cross multiply to determine if ratios are equal.

TO THE CLASS	ON THE BOARD
Suppose there are 6 basketballs and 2 have stripes.	(image of 6 basketballs, 2 striped)
What is the ratio of striped basketballs to the total number of basketballs? (2 out of 6 are striped.)	$\frac{2 \text{ striped basketball}}{6 \text{ basketballs}}$
Can this be simplified? (yes)	$\frac{1 \text{ striped basketball}}{3 \text{ basketballs}}$
Cross multiply by multiplying the numerator of one ratio by the denominator of the other ratio.	$6 \times 1 = 6$ $\frac{2}{6} \bowtie \frac{1}{3}$ $2 \times 3 = 6$
If you cross multiply these ratios, are the products the same? (yes)	$6 = 6$
Since the products are equal, then the two ratios are equal. Equal ratios form a proportion.	$\frac{2}{6} = \frac{1}{3}$ is a proportion.

EXAMPLE On a test, Suki got 6 answers right out of 9. Jim got 8 answers right out of 10. Are the students' scores proportional? (**No**)

ASK	ANSWER
How should you set up the ratios for these two students?	$\frac{6}{9}$ $\frac{8}{10}$
What are the cross products for these two ratios?	$6 \times 10 = 60$ and $9 \times 8 = 72$
Compare the cross products. Are these two ratios equal?	No. $60 \neq 72$
Are the student's scores proportional? Explain.	No, because a proportion is a statement of equal ratios and these two ratios are not equal.

17 Practice Answers

Complete solutions to all Practice problems on student page 17 are at the back of this book on p. 39.

1. $\frac{2}{3} = \frac{12}{18}$ 2. $\frac{3}{8} \neq \frac{6}{18}$
3. $\frac{2}{5} = \frac{8}{20}$ 4. $6:8 = 12:16$
5. $4:9 \neq 12:26$
6. $\frac{3}{7} \neq \frac{14}{6}$ 7. $\frac{7}{2} = \frac{21}{6}$
8. $9:5 \neq 27:9$
9. $\frac{2}{8} = \frac{20}{80}$
10. $\frac{40}{4} = \frac{60}{6}$; yes
11. $\frac{3}{10} \neq \frac{5}{20}$; no
12. $\frac{6}{15} \neq \frac{9}{21}$; no
13. $\frac{50}{1} = \frac{200}{4}$; yes
14. $\frac{0.5}{10} \neq \frac{1}{25}$; no
15. $\frac{8}{10} \neq \frac{15}{20}$; no

On Your Own

This summer 8 out of 12 people took a vacation. Last summer 3 out of 4 people took a vacation. Are these ratios proportional? (**No**)

- Ask, *What are the ratios for this problem?* ($\frac{8}{12}$ and $\frac{3}{4}$)

- Ask, *Are the cross products equal?* (**no, 32 ≠ 36**)

Direct students to **Practice Proportions** on page 17. Have them work in groups and correct each other's answers.

Using Cross Products Student pages 18 19 20

LESSON 5

18 Students learn how to use cross products to find a missing value in a proportion. Say, *You can use cross products to decide if two ratios are equal and form a proportion.*

TO THE CLASS	ON THE BOARD
If a value is missing in a proportion, as this example shows, how can you find the missing value?	$\frac{12}{3} = \frac{?}{120}$
Try using cross multiplication. What equation could you write? Let *n* be the missing value.	$\frac{12}{3} = \frac{n}{120}$ $3 \times n = 1440$
How can you solve for the missing value? **(divide by 3)**	$\frac{1440}{3} = n = 480$
What value is needed to make the ratios a proportion?	480

EXAMPLE *Keisha got 4 hits in 12 at-bats. If Keisha keeps getting hits at this same rate, how many hits will she get in 60 at-bats?* **(20)**

TO THE CLASS	ON THE BOARD
How do you set up a proportion using the information given?	$\frac{4}{12} = \frac{n}{60}$
What do you need to cross multiply to solve the proportion?	4×60 and $12 \times n$
Write an equation to solve for *n*.	$4 \times 60 = 12 \times n$ $12n = 240$ $\frac{12n}{12} = \frac{240}{12}; n = 20$
How many hits would Keisha get in 60 at-bats?	Keisha would get 20 hits in 60 at-bats.

On Your Own

One photo is 6 inches long and 4 inches wide. You can enlarge the photo so that it will be 18 inches long. How wide will the photo be? **(12 inches)**

- Ask, *What is missing for the second photo?* **(the width of the enlarged photo)**
- Guide students to set up a proportion. ($\frac{6}{4} = \frac{18}{n}$)
- Say, *Do cross multiplication and write an equation to solve for* n. ($6n = 72$, $n = 12$)

Direct students to **Practice Using Cross Products** on page 19. Have them work in groups and correct each other's answers.

LEARNING STYLES

VISUAL Have students use a colored pencil to show division on both sides of the equation. This will help students remember this step in the process as they solve for the variable.

Practice Answers 19

Complete solutions to all Practice problems on student page 19 are at the back of this book on p. 40.

1. $9 = n$ 2. $n = 6$
3. $2 = n$ 4. $25 = x$
5. $6 = n$ 6. $n = 25$
7. $x = 20$ 8. $y = 56$
9. $y = 55$ 10. $90
11. 38 rolls 12. 14 laps
13. 210 minutes   14. 320 points
15. $108

Test-Taking Strategy 20

WRITE PROPORTIONS TO SOLVE WORD PROBLEMS
You can write proportions to solve word problems.

EXAMPLE
Ask for volunteers to read aloud the steps.

TRY IT OUT
Have students work out the problem on their own, using a proportion to solve. Review how to set up equal ratios to find the correct answer. (Hint: *It may be easier to reduce $\frac{10}{15}$ first.*)
Answer: C

LESSONS 6–7 OVERVIEW

Skills

Write proportions for everyday situations.

Use equivalent ratios and proportional reasoning to solve problems.

KEY WORDS

proportion: a statement that says two ratios are equal

equivalent: two things are equal

proportional reasoning: using proportions as a way to reason and solve everyday problems

term: the name given to each value in a proportion

cross products: the answer to multiplying the numerator of one ratio by the denominator of the other ratio in a proportion

LEARNING TIPS

- To check if answers are correct, replace the variable in the proportion and cross multiply.
- Label the units in each term of the proportion to ensure that it is set up correctly.
  Example: $\frac{miles}{gallons} = \frac{miles}{gallons}$

English Language Learners

- To help students understand variables, have the letter represent the first initial in the missing term. For example, if the missing term is crackers, use the variable c to represent crackers.

More on Proportions   Student page 21

Access Prior Knowledge

If you buy 10 gallons of gasoline for $18, how much will 20 gallons of gasoline cost? **Cross multiply to find the missing value. $36**

Guided Instruction

Direct students to the map. Have students read the overview.

ASK	ANSWER
Look at the map. What ratio is shown on the scale to help you figure out the distance from one place to another on the map?	$\frac{1}{2}$ inch = 15 miles, or 0.5 inches = 15 miles
Use a ruler to measure the length from Highmount to Allaben on the map. What is the length?	2 inches
Using this information, what proportion can you write to find the mileage between Highmount and Allaben?	$\frac{0.5}{15} = \frac{2}{n}$
Solve the proportion. What is the distance from Highmount to Allaben?	60 miles

Remember the Basics

Direct students to the paired ratios shown at the bottom of the page. In this activity, students are to use cross products to solve each proportion. If students are having difficulty, review finding cross products.

How can you tell if two ratios are equal? (**The products from cross multiplying will be equal.**)

$\frac{3}{4} = \frac{n}{12}$	$\frac{30}{18} = \frac{10}{n}$	$\frac{3}{8} = \frac{n}{40}$
$3 \times 12 = 4 \times n$	$30 \times n = 18 \times 10$	$3 \times 40 = 8 \times n$
$4n = 36$	$30n = 180$	$120 = 8n$
$n = 9$	**$n = 6$**	**$n = 15$**

Writing Proportions

LESSON 6

Student pages 22 23

22 Students learn how to write and solve proportions.

TO THE CLASS	ON THE BOARD
The values in a proportion are called *terms*. How many terms are in a proportion?	4 terms: $\dfrac{\text{term}}{\text{term}} = \dfrac{\text{term}}{\text{term}}$
How could you write a proportion to show 5 servings in 1 can and 10 servings in 2 cans? Label the terms and the values.	$\dfrac{5 \text{ servings}}{1 \text{ can}} = \dfrac{10 \text{ servings}}{2 \text{ cans}}$ $\dfrac{\text{servings}}{\text{can}} = \dfrac{\text{servings}}{\text{can}}$
Use cross products to solve this proportion. Are the ratios equal?	$\dfrac{5}{1} = \dfrac{10}{2}; 5 \times 2 = 10 \times 1$ $10 = 10$ Yes, the ratios are equal.

LEARNING STYLES

VISUAL Have students use two different-colored markers to differentiate between the types of terms in the ratios. Have them use one color for all terms that are numerators and the second color for the terms that are denominators. The students will see that cross multiplying will always create an equation with the two colors on both sides.

EXAMPLE *Use the nutrition facts from the picture of the label. The nutrition label states that there are 70 calories in 2 crackers. You eat 7 crackers. How many calories are in 7 crackers?* (**245 calories**)

ASK	ANSWER
What information is given in the problem that will help you solve the problem?	70 calories in 2 crackers 7 crackers eaten
What information do you need to know to solve this problem?	The number of calories in 7 crackers.
What proportion can you write to identify the 4 terms?	$\dfrac{\text{calories}}{\text{crackers}} = \dfrac{\text{calories}}{\text{crackers}}$
What values should you write to complete these terms?	$\dfrac{70 \text{ calories}}{2 \text{ crackers}} = \dfrac{n \text{ calories}}{7 \text{ crackers}}$ or, $\dfrac{70}{2} = \dfrac{n}{7}$
What do you solve for *n*?	$2n = 70 \times 7; n = 245$
How many calories are in 7 crackers?	There are 245 calories in 7 crackers.

On Your Own

Julio can type 90 words in 2 minutes. He needs to take a 5-minute typing test in order to get a job. If Julio types at his usual rate, how many words will he type on this test? (**225 words**)

- Ask students to identify the question and data in the problem. (**How many words will Julio type on the test? 90 words for 2 minutes; a 5-minute test.**)

- Direct students to write a proportion and solve for the missing value. ($\dfrac{90}{2} = \dfrac{n}{5}; n = 225$)

- Invite a volunteer to write the proportion on the board and explain how he or she solved the problem.

Direct students to **Practice Writing Proportions** on page 23. Have them work in groups and correct each other's answers.

Practice Answers

23

Complete solutions to all Practice problems on student page 23 are at the back of this book on p. 41.

1. $\dfrac{3}{2} = \dfrac{n}{6}$ 2. $\dfrac{42}{3} = \dfrac{147}{n}$
3. $\dfrac{20}{5} = \dfrac{n}{30}$ 4. $\dfrac{7}{14} = \dfrac{n}{8}$
5. $\dfrac{18}{1} = \dfrac{n}{9}$ 6. $\dfrac{40}{5} = \dfrac{n}{40}$
7. $\dfrac{171}{3} = \dfrac{n}{1}$ 8. $\dfrac{2}{7} = \dfrac{30}{n}$
9. $\dfrac{\$390}{30} = \dfrac{n}{1}; n = \$13$
10. $\dfrac{360}{6} = \dfrac{n}{1}; n = 60$ calories
11. $\dfrac{8}{20} = \dfrac{n}{60}; n = 24$ laps
12. $\dfrac{8}{12} = \dfrac{n}{30}; n = 20$ calls
13. $\dfrac{36}{1} = \dfrac{n}{0.5}; n = 18$ math problems
14. $\dfrac{2}{12} = \dfrac{0.5}{n}; n = 3$ servings

LESSON 7

Problem Solving Using Proportions
Student pages 24 25 26

LEARNING STYLES

AUDITORY As students form ratios to create a proportion, tell them to say the units out loud. Listening as they say the units helps students check that they have written the same units in the same order.

24 Students learn how to use proportions to solve problems.

TO THE CLASS	ON THE BOARD
When comparing ratios, does it matter if the units are placed in the same order? **(yes)**	$\dfrac{2 \text{ miles}}{30 \text{ minutes}} = \dfrac{5 \text{ miles}}{75 \text{ minutes}}$
You can use proportions to solve everyday problems. For example, if there are 60 minutes in an hour, how many minutes are in 4 hours?	$\dfrac{60 \text{ minutes}}{1 \text{ hour}} = \dfrac{n \text{ minutes}}{4 \text{ hours}}$
If proportions involve rates, do the rates need to be the same? **(Yes, rates stay the same.)**	$\dfrac{\text{miles}}{\text{hour}} = \dfrac{\text{miles}}{\text{hour}}$

EXAMPLE You are making blueberry muffins. You have $15 to buy blueberries. How many packages of blueberries can you buy? **(9 pints)**

ASK	ANSWER
What question are you asked to solve?	How many packages of blueberries can you buy?
What information will help you answer this question?	3 pints of blueberries cost $5. You have $15 to spend.
What proportion can you write to show this comparison?	$\dfrac{3 \text{ pints blueberries}}{\$5} = \dfrac{n \text{ pints blueberries}}{\$15}$
Are the same units used in the same order?	yes; $\dfrac{\text{blueberries}}{\text{price}} = \dfrac{\text{blueberries}}{\text{price}}$
How would you solve this proportion?	$\dfrac{3}{5} = \dfrac{n}{15}$ $5n = 45$ $n = 9$
How many packages of blueberries can you buy?	9 pints of blueberries

25 **Practice Answers**

Complete solutions to all Practice problems on student page 25 are at the back of this book on p. 41.

1. $n = 4$; 4 pounds
2. $n = 12$; $12
3. $n = 4$; $4
4. $n = 9$; 9 roses
5. $n = 105$; 105 miles
6. $n = 1\frac{1}{8}$; $1\frac{1}{8}$ cups of water
7. $n = 6$; 6 calls
8. $n = 0.5$; 0.5 pound
9. $n = 2.5$; 2.5 songs
10. $n = 40$; 40 black marbles

26 **Test-Taking Strategy**

WRITE AN EQUATION
You can write an equation to solve proportions.

EXAMPLE
Ask for volunteers to read aloud the test-taking strategy, the example problem, and the steps.

TRY IT OUT
Have students work out the problem on their own by writing an equation to solve the proportion. Review how to set up equal ratios to find the correct answer. **Answer: D**

On Your Own

Look at the ad again. You need to buy some honeydew melons. How much will 5 melons cost? **($12.50)**

- Ask, *What information can you get from the ad?* **(the cost of 2 melons)**
- Guide students to create a proportion. ($\dfrac{2 \text{ melons}}{\$5.00} = \dfrac{5 \text{ melons}}{n}$)
- Tell students to solve the proportion. ($\dfrac{2}{5.00} = \dfrac{5}{n}$; $2n = 25.00$; $n = 12.50$)
- Invite a volunteer to write the proportion on the board and explain how he or she solved the problem.

Direct students to **Practice Problem Solving Using Proportions** on page 25. Have them work in groups and correct each other's answers.

Review Test-Taking Strategies

Student pages 14 20 26

Ask for volunteers to name a Test-Taking Strategy they learned in this unit. Have them explain how to use the strategy, and why they find it helpful.

USE LOGICAL REASONING — 14

You can draw and label Venn diagrams to solve ratio problems.

WRITE PROPORTIONS TO SOLVE WORD PROBLEMS — 20

You can write proportions for equal ratios to solve word problems.

WRITE AN EQUATION — 26

You can write an equation for a proportion to solve word problems.

USE A STRATEGY

- Divide students into small groups and assign them one of the strategies in the unit.
- Using the model from their books, have each group create 3 new problems. Have the group solve each problem on a separate sheet of paper.
- Invite groups to exchange and solve problems.
- After groups have finished solving problems, they can exchange their solutions.

Mixed-Level Classroom

Student pages 7 15 21

Remind students of the short-term goals they wrote at the beginning of this book on page 3, *Setting Goals*. Say, *Each learning objective is a short-term goal, a stepping-stone that leads you to your long-term goal.*

OVERVIEW	LEARNING OBJECTIVES	PAGE
Lessons 1–3	• Understand that ratios compare two quantities. • Reduce ratios to simplest form. • Understand rates and unit rates.	7
Lessons 4–5	• Set up a proportion. • Use cross products to solve proportions.	15
Lessons 6–7	• Write proportions for everyday situations. • Use equivalent ratios and proportional reasoning to solve problems.	21

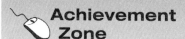

Achievement Zone

You can download reproducible worksheets that reinforce ratios, rates, and proportions at www.HarcourtAchieve.com/AchievementZone.

RETEACH

Write these problems on the board and invite students to solve them.

1. Write a ratio for the number of days in a week to the number of days in a year. (**7 to 365**)

2. Write this rate in simplest form: $20 for 10 gallons of gasoline. ($\frac{20}{10} = \frac{2}{1}$)

REMIND STUDENTS

- Always make sure that the terms in each ratio are organized.
- In a proportion, you need to have the same units in the same order.
- Use cross products as a way to check your solution to a proportion.

UNIT REVIEW ANSWERS — 27

1. 2 : 3   2. 3 : 7   3. 2 to 3   4. 3 to 6   5. 7 grams of fat in one serving
6. 1.5 pages per minute, 90 pages per hour   7. No   8. Yes
9. No; Adam's rate of $18 per hour is higher.   10. Yes   11. $n = 72$
12. $n = 15$ 13. $n = 4$ 14. $\frac{5}{7.5} = \frac{n}{30}$; $n = 20$ games
15. $\frac{24}{8} = \frac{n}{5}$; $n = 15$ tennis balls

UNIT 2

Fractions, Decimals, and Percents Student page 28

Students explore fractions, decimals, and percents by considering the number of registered voters who turn out to vote in an election.

Learning Objectives

In Unit 2, students will

- Convert between fractions, and percents and decimals and percents.
- Use the percent triangle to find the part, percent, and whole.
- Write a percent equation.

Real-Life Matters

Invite volunteers to read the Real-Life Matters on page 28. Say, *You can write 51% as a fraction. Fifty-one percent is neither a high percentage nor a low percentage. In fact, it is about half.*

TO THE CLASS	ON THE BOARD
This ratio compares the people who voted to the number of people who are registered to vote. For every 100 registered voters, 51 actually voted.	$51\% = \frac{51}{100}$ $\frac{voted}{registered} = \frac{51}{100}$
If you were to express this ratio as a fraction, you could say about half the people who were registered to vote actually voted.	$\frac{51}{100}$ almost equals $\frac{50}{100}$ $\frac{50 \div 50}{100 \div 50} = \frac{1}{2}$

LEARNING TIPS

- Percents are fractions with a denominator of 100.
- Percent means *per hundred*.

Real-Life Application

Let students work together to answer the questions presented.

- Say, *75% means $\frac{75}{100}$. What is this fraction in its simplest form?* ($\frac{75}{100} = \frac{3}{4}$)

- Ask, *Is 75% a high percentage? How do you know?* (**Possible answers: Yes, 75% is like having 75 cents, or 3 quarters, and that is close to a whole dollar; in this Real-Life Application, 75% means 3 out of 4 students.**)

- Ask, *How could you use proportions to find out how many students you need to get out and vote?* (**Let v stand for voters: $\frac{3}{4} = \frac{v}{940}$. Cross-multiply: $4v = 3 \times 940$; $v = 705$.**)

- Ask, *Do you know an easy way to change a percent to a decimal?* (**Invite students to explain how to write 75% as a decimal to someone who does not know what fractions or decimals are.**)

FOLLOW-UP

Would you rather have 12% of 200 CDs or 25% of 76 CDs? **(24 CDs or 19 CDs)**

OVERVIEW LESSONS 8–10

Fractions and Percents Student page 29

Access Prior Knowledge

How can you write the statement 2,000 out of 2,500 high school students are right-handed *in mathematical terms?* **Use a ratio:**

$$\frac{2{,}000 \text{ right-handed students}}{2{,}500 \text{ total students}} \text{ or } \frac{2{,}000}{2{,}500}$$

Guided Instruction

Direct students to the Overview instruction that discusses election candidates and votes.

TO THE CLASS	ON THE BOARD
Suppose 2 out of 5 candidates for mayor are men. How could you use a ratio to describe this comparison?	2 out of 5 or $\frac{2}{5}$ are men
If the other 3 candidates are women, then 3 out of 5 candidates are women. How can you write this ratio?	3 out of 5 or $\frac{3}{5}$ are women
What fractions show the comparisons? Fractions work well in these situations since the size of the group is small.	$\frac{2}{5}$ and $\frac{3}{5}$
What if 11,730 of the 20,400 votes went for your favorite candidate. What fraction shows this comparison? Does this fraction show the information clearly? (**No, the large numbers make the information harder to understand.**)	$\frac{11{,}730}{20{,}400}$
How could you write this fraction as a percent? If you express this fraction as a percent, you see that about 58% or 58 out of every 100 voters voted for your favorite candidate.	$\frac{11{,}730}{20{,}400} = 57.5\%$

Remember the Basics

Direct students to the chart at the bottom of the page. In this activity, students fill in the empty spaces on the chart, rewriting fractions and decimals as fractions in lowest terms. Say, *You need to know how to change decimals to fractions to solve percent problems. How do you change a decimal to a fraction?* (**Read the decimal aloud. Write the decimal as a number over a multiple of ten.**)

1. 4 out of 12 = $\frac{1}{3}$	2. 60 out of 100 = $\frac{3}{5}$	3. 8 out of 20 = $\frac{2}{5}$
4. 0.75 = $\frac{3}{4}$	5. 0.55 = $\frac{11}{20}$	6. 0.07 = $\frac{7}{100}$

Skills

Learn that percents and fractions show a ratio out of 100.

Use proportions to write fractions as percents.

Write percents as fractions.

KEY WORDS

percent: a special ratio that compares a number to 100

LEARNING TIPS

- Remember when converting percents to a ratio in fraction form that percent means *per hundred,* or *for every hundred*.
- When comparing large numbers, a percent gives you a better representation than a fraction.

English Language Learners

- Learning to recognize the meaning of common base words can help students build their vocabulary. Some students may benefit from looking up words in the dictionary.
- Pair fluent English language speakers with ELLs and have them research the meaning of word roots in *fraction*, *decimal*, and *percent*. Later, students can make connections to other words with the same word base, such as *fracture*, *decade*, *century*.

LESSON 8

Understanding Percents

Student pages 30 31

LEARNING STYLES

KINESTHETIC Have students draw and cut out models to represent percents. First have them draw and cut out a circle, shade $\frac{1}{4}$ of it, and write 25% in it. Next have them draw and cut out another circle, shade half of it, and write 50% in it. Then have them draw and cut out another circle, shade all of it, and write 100% in it. Finally have them draw and cut out a circle without shading any of it, then write 0% in it.

30 Students learn how percents and ratios are related.

TO THE CLASS	ON THE BOARD
What did Mia pay for a dog collar?	39¢
If there are 100 cents in a dollar, what fraction shows 39¢?	$\frac{39}{100}$
The fraction $\frac{39}{100}$ also is a ratio. Since percent means per hundred, what is the ratio as a percent?	$\frac{39}{100} = 39\%$
What are some other money values that can be written as a percent of a dollar?	Dime – $\frac{10}{100}$ – 10% Quarter – $\frac{25}{100}$ – 25% Dollar – $\frac{100}{100}$ – 100%

EXAMPLE You pay $100 to have your car repaired. You pay $15 more for the needed car part. The cost of the part is what percent of the repair bill? (**15%**)

31 Practice Answers

Complete solutions to all Practice problems on student page 31 are at the back of this book on p. 43.

1. 44% 2. 58% 3. 91%
4. 30% 5. 75% 6. 85%
7. 66% 8. 92% 9. 53%
10. 14% 11. 50% 12. 17%
13. 23% 14. 46%

ASK	ANSWER
What information do you know from the problem?	$100 is the cost of the total repair. $15 is the cost of the part.
What ratio can you write that compares the cost of the car part to the total repair bill?	$\frac{\$15}{\$100}$ or $\frac{15}{100}$
How would you write this ratio as a percent?	$\frac{15}{100} = 15\%$
The cost of the part is what percent of the repair bill?	15% of the total bill

On Your Own

One hundred students tried out for the school play. Twenty-two of the students got a part in the play. Write this ratio as a percent. (**22%**)

- Ask, *What information is given in the problem?* (**100 students tried out for the play; 22 got parts.**)

- Guide students to write a ratio for students who get parts in the play to the number of students who tried out. ($\frac{22}{100}$)

- Tell students to change the ratio to a percent. ($\frac{22}{100} = 22\%$)

- Invite a volunteer to write the ratio on the board and explain how to change the ratio to a percent.

Direct students to **Practice Understanding Percents** on page 31. Have them work in groups and correct each other's answers.

Converting Fractions to Percents Student pages 32 33

LESSON 9

32 Students learn how to convert fractions to percents.

TO THE CLASS	ON THE BOARD
There are several different ways that you can change a fraction to a percent. One way is to multiply the fraction by 100%. How can you change $\frac{1}{5}$ to a percent by multiplying by 100?	$\frac{1}{5} \times 100 = \frac{1}{5} \times \frac{100}{1}$ $\frac{1}{5} \times \frac{100}{1} = \frac{20}{1} = 20\%$
Another way to change a fraction to a percent is to divide. How do you write $\frac{1}{5}$ as a division problem?	$\frac{1}{5} = 1 \div 5$ $5 \overline{)1}$
What do you get when you divide?	0.20
How do you write the decimal as a percent? (**Move the decimal point two places to the right and write the percent sign.**)	$0.20 = 20\%$

EXAMPLE *Ahmad got 6 answers correct out of 8 questions on a math quiz. What percent did Ahmad get correct?* (**75%**)

TO THE CLASS	ON THE BOARD
What fraction can be written using the information given?	$\frac{6}{8}$
How do you find the percent by multiplying by 100?	$\frac{6}{8} \times 100 = \frac{3}{4} \times 100$ $\frac{3}{4} \times \frac{100}{1} = \frac{75}{1} = 75\%$
How do you write $\frac{6}{8}$ as a division problem?	$\frac{6}{8} = 6 \div 8 = 8\overline{)6}$
What percent did Ahmad get correct?	Ahmad got 75% correct.

On Your Own

Germaine played a video game. In the game, she hit 12 of the 20 targets. What percentage of the targets did she hit? (**60%**)

- Guide students to write a fraction and reduce it to lowest terms. ($\frac{12}{20} = \frac{3}{5}$)
- Tell students to decide which operation they will use to solve the problem. ($\frac{3}{5} \times 100$ or $3 \div 5 \times 100$)
- Invite a volunteer to write the fraction on the board and explain how he or she would calculate the percent. ($\frac{3}{5} \times 100 = 60\%$)

Direct students to **Practice Converting Fractions to Percents** on page 33. Have them work in groups and correct each other's answers.

LEARNING STYLES

VISUAL Have students make a chart of all the factors of 100. Have them list all combinations that make 100. Tell them to use their factor chart when looking for a number to use to change a denominator to 100.

Math Toolkit 78

Use the Equivalent Fractions, Decimals, and Percents chart on student page 78 to help students understand the relationship between fractions and percents.

Practice Answers 33

Complete solutions to all Practice problems on student page 33 are at the back of this book on p. 43.

1. 80%   2. 90%   3. 55%
4. 6%   5. 37.5% 6. 22%
7. 32%   8. 70%   9. 20%
10. 85% 11. 80%   12. 70%
13. 87.5% 14. 6%   15. 20%

LESSON 10

Converting Percents to Fractions
Student pages 34 35 36

LEARNING STYLES

AUDITORY Have students listen for the word *percent* as they read percentages. Remind them that percent means *per hundred*. When students say *percent*, they should think of a fraction with a denominator of 100.

78 Math Toolkit

Use the Equivalent Fractions, Decimals, and Percents chart on student page 78 to help students understand the relationship between fractions and percents.

35 Practice Answers

Complete solutions to all Practice problems on student page 35 are at the back of this book on p. 43.

1. $\frac{11}{20}$ 2. $\frac{19}{20}$ 3. $\frac{3}{20}$ 4. $\frac{11}{25}$
5. $\frac{83}{100}$ 6. $\frac{7}{25}$ 7. $\frac{2}{25}$ 8. $\frac{17}{50}$
9. $\frac{17}{25}$ 10. $\frac{9}{20}$ 11. $\frac{43}{100}$ 12. $\frac{3}{20}$
13. $\frac{17}{25}$ 14. $\frac{29}{100}$ 15. $\frac{7}{25}$

36 Test-Taking Strategy

USE A GRAPH
You can use a circle graph to help you answer questions about percents.

EXAMPLE
Ask for volunteers to read aloud the example and the steps.

TRY IT OUT
Have students work out the problem on their own using the circle graph. Review how to find the correct answer. **Answer: C**

---

34 Students learn to convert percents to fractions.

TO THE CLASS	ON THE BOARD
What does percent mean?	Percent → per hundred Example A: $33\% = \frac{33}{100}$
How can you write 33% as a fraction with a denominator of 100? Can $\frac{33}{100}$ be simplified? **(No, it is in lowest terms.)** So 33% as a fraction in lowest terms is $\frac{33}{100}$.	$33\% = \frac{33}{100}$
What about 25% as a fraction? Can it be simplified when converted to a fraction? **(Yes, the final answer can be reduced.)**	Example B: $25\% = \frac{25}{100}$ $\frac{25}{100} \div \frac{25}{25} = \frac{1}{4}$

EXAMPLE *Thirty percent of the students attend a driver's ed class after school. What fraction of students attend driver's ed after school?* ($\frac{3}{10}$)

ASK	ANSWER
What percent is stated in this problem?	30%
How would you write 30% as a fraction?	$\frac{30}{100}$
Is this fraction in simplest form?	No.
What is the GCF of 30 and 100?	10
If you divide the numerator and denominator by 10, what is the equivalent fraction for $\frac{30}{100}$?	$\frac{30 \div 10}{100 \div 10} = \frac{3}{10}$
What fraction of the students attend driver's ed after school?	$\frac{3}{10}$ of the students

On Your Own

In this year's graduating class, 85% of the students plan on going to college. What fraction of the students plan to go to college? ($\frac{17}{20}$)

- Direct students to write the percent as a fraction. ($85\% = \frac{85}{100}$)

- Ask, *Is the fraction in lowest terms?* (**no**)

- Invite a volunteer to write the fraction on the board, determine if the fraction is in lowest terms, and explain how he or she arrived at the solution. (**Reduce the fraction $\frac{85}{100}$ to $\frac{17}{20}$ using 5 as the GCF.**)

Direct students to **Practice Converting Percents to Fractions** on page 35. Have them work in groups and correct each other's answers.

18 LESSON 10

OVERVIEW LESSONS 11–12

Decimals and Percents   Student page 37

Access Prior Knowledge

When you hear that there is a 90% chance of rain, what does that mean? **There is a 90 out of 100, or 9 out of 10, chance that it will rain.**

Guided Instruction

Direct students to the Overview instruction that displays a lunch bill of $25.

ASK	ANSWER
How much is the lunch bill?	$25
You want to leave an 18% tip. How would you write 18% of $25?	18% of $25
Is there another way to write 18% of $25? With percents, *of* means multiplication.	18% × $25
What are three different ways to write 18%?	Percent = Fraction = Decimal 18% = $\frac{18}{100}$ = 0.18
Which way of writing 18% will make it easy to figure out the tip on a $25 bill?	As a decimal: 0.18 0.18 × 25
What is 18% of $25?	0.18 × $25 = $4.50

Remember the Basics

Direct students to the chart at the bottom of the page. In this activity, students rewrite percents as fractions in lowest terms or rewrite fractions as percents. Say, *When working with percents, you need to know how to change a fraction to lowest terms. How can you tell when you have expressed a percent as a fraction in lowest terms?* **(when the GCF for the numerator and denominator is 1)**

1. $\frac{3}{12} = \frac{n}{100} = 25\%$
   $12n = 300$
   $n = 25$
   $\frac{3}{12} = 25\%$

2. $\frac{6}{15} = \frac{n}{100}$
   $15n = 600$
   $n = 40$
   $\frac{6}{15} = 40\%$

3. $\frac{8}{40} = \frac{n}{100}$
   $40n = 800$
   $n = 20$
   $n = 20\%$

4. $72\% = \frac{72}{100} = \frac{18}{25}$

5. $45\% = \frac{45}{100} = \frac{9}{20}$

6. $36\% = \frac{36}{100} = \frac{9}{25}$

Skills

Write any decimal as a percent.
Write any percent as a decimal.

KEY WORDS

percent: per hundred

LEARNING TIPS

- Change percents to fractions by using 100 as the denominator.
- Multiplying decimals by 100 moves the decimal point two places to the right, and shows the percent that the decimal is equal to. For example, 0.08 × 100 = 8, so 0.08 = 8%.

English Language Learners

- Remind students to work on their own personal glossary daily. They can list words they do not know or any words with new meanings. For example, *compute, express, extend, favor, figure, lounge,* and *survey* are all words that ELLs should know.

LESSON 11

Converting Decimals to Percents Student pages 38 39

38 Students learn how to convert decimals to percents.

LEARNING STYLES

KINESTHETIC Have students cut a small circle out of colored self-sticking paper. Next, have students write a number on centimeter graph paper. Use the circle for the decimal point. Finally, have the students change the decimal to a percent by physically moving the circle two places to the right.

78 Math Toolkit

Use the Equivalent Fractions, Decimals, and Percents chart on student page 78 to help students understand the relationship between decimals and percents.

TO THE CLASS	ON THE BOARD
74 hundredths of the high school students surveyed want a new student lounge. How do you write this value as a decimal?	0.74
What should you multiply by to write 0.74 as a percent? (Multiply by 100. Notice how the decimal place moved two places to the right.)	$0.74 \times 100 = 74$
What is 0.74 written as a percent?	74%
How many students want a new lounge?	74% of the students surveyed want a new lounge.

EXAMPLE *A customer calculated that 0.294 people in a grocery store buy green beans. What percent of people in the store buy green beans?* (**29.4%**)

TO THE CLASS	ON THE BOARD
To write 0.294 as a percent, what must you use to multiply? (Multiply by 100.)	$0.294 \times 100 = 29.4$
What happens to the decimal?	Draw arrow to show decimal moving 2 places to the right.
What percent of people in the store buy green beans?	29.4%

39 Practice Answers

Complete solutions to all Practice problems on student page 39 are at the back of this book on p. 43.

1. 6%	2. 64%
3. 5.2%	4. 88.8%
5. 70%	6. 0.5%
7. 0.24%	8. 90.8%
9. 140%	10. 106.3%
11. 45%	12. 95%
13. 35%	14. 5%
15. 34.8%	16. 98%

On Your Own

According to a survey, 0.8 of the students in Neela's school like the idea of extending the school day to 5:00 PM. What percent of students want to extend the school day? (**80%**)

- Guide students to multiply 0.8 by 100. ($0.8 \times 100 = 80$)
- Ask, *What happened to the decimal point?* (**The decimal point moved two places to the right.**)
- Tell students to write 0.8 as a percent. (**80%**)
- Invite a volunteer to write the problem on the board and explain how he or she arrived at the solution.

Direct students to **Practice Converting Decimals to Percents** on page 39. Have them work in groups and correct each other's answers.

Converting Percents to Decimals

Student pages 40 41 42

LESSON 12

40 Students learn to convert percents to decimals.

TO THE CLASS	ON THE BOARD
How do you write 35 hundredths as a percent? (**To write a decimal as a percent, first multiply by 100 and then write the percent sign.**) Notice that the decimal point moved two places *to the right*.	decimal → percent 0.35 = ____ % 0.35 × 100 = 35 0.35 = 35%
What do you do to change a percent to a decimal? Remove the percent sign and divide by 100. Notice that the decimal point moved two places *to the left*.	percent → decimal 35% = 35 ÷ 100 = 0.35
What is 48% as a decimal?	48% → 48 48 ÷ 100 = 0.48

EXAMPLE Only 6% of teens questioned said that watching movies at home is better than seeing them in the theater. What is 6% written as a decimal? (**0.06**)

ASK	ANSWER
To write 6% as a decimal, what do you have to remove first? Notice the placement of the decimal point.	Remove the percent sign. 6% → 6
Percent means per hundred, so what number should you divide 6 by? Notice that moving the decimal point two places to the left resulted in adding a zero as a place holder.	Divide by 100. 6 ÷ 100 = .06 Move the decimal two places to the left.
Should you add a zero in the ones place?	Yes: 0.06
What is 6% written as a decimal?	0.06

On Your Own

The newspaper reported that 25.5% of people surveyed watched the news every night. How would you write this as a decimal? (**0.255**)

- Direct students to remove the percent sign and note the location of the decimal point. (**25.5**)
- Ask, *Percent means* per hundred. *What number do you use to divide?* (**100**)
- Tell students to divide and find the decimal. (**0.255**)
- Ask, *What did you notice about the decimal point?* (**It moved two places to the left.**)
- Invite a volunteer to change the percent to a decimal on the board and explain how he or she arrived at the solution.

Direct students to **Practice Converting Percents to Decimals** on page 41. Have them work in groups and correct each other's answers.

LEARNING STYLES

AUDITORY Have students say aloud the percent number and the decimal name. Tell them to notice that the endings are different. The different endings indicate that the decimal point has moved.

Practice Answers 41

Complete solutions to all Practice problems on student page 41 are at the back of this book on p. 44.

1. 0.85 2. 0.65
3. 0.12 4. 0.48
5. 0.73 6. 0.29
7. 0.04 8. 0.347
9. 1.50 10. 0.025
11. 0.031 12. 0.003
13. 0.74 teens
14. 0.40 moviegoers
15. 0.48 teens
16. 0.05 teens
17. 0.375 movies
18. 1.20 effort

Test-Taking Strategy 42

DRAW A NUMBER LINE
You can use a number line to help you answer questions about percents.

EXAMPLE
Ask for volunteers to read the example and the steps aloud.

TRY IT OUT
Have students work out the problem on their own using the number line. Review how to find the correct answer. **Answer: B**

LESSONS 13–15 OVERVIEW

Skills

Solve for the whole, the part, and the percent.

Use a percent triangle to solve percent problems.

KEY WORDS

whole: the entire amount or quantity or original amount

part: a portion of a whole

percent: a special ratio that compares a number to 100

equivalent: equal

LEARNING TIPS

- Simplify a fraction before changing it to a decimal.
- Remember to multiply the decimal by 100 to a percent.
- The zero to the left of the decimal point acts as a placeholder.

English Language Learners

- Write *whole* and *hole* on the board. Point out that *whole* is spelled with a *w*. Discuss the meanings of *whole* and *hole*.
- Make sure that students understand the meaning of *equivalent* as *equal* or *having the same value*.

Percents and Problem Solving I Student page 43

Access Prior Knowledge

Ask, *How do you change a decimal to a percent?* **To change the decimal to a percent, multiply the decimal by 100 and put on the percent sign.**

Guided Instruction

Direct students to the pizza illustration.

TO THE CLASS	ON THE BOARD
Percent problems have three pieces of information that you need to know: the part, the whole, and the percent.	part whole percent
How many slices of pizza are there in the part?	part = 2
How many slices of pizza are in the whole pizza? **(8)** You can say that you have 2 pieces out of 8 pieces.	whole = 8
How can you show 2 out of 8? ($\frac{2}{8}$) What is $\frac{2}{8}$ in simplest form? ($\frac{1}{4}$)	$\frac{2}{8} = \frac{1}{4}$
How can you write $\frac{1}{4}$ as a decimal? **(0.25)**	$\frac{2}{8} = \frac{1}{4} = 0.25$
How would you write 0.25 as a percent?	$0.25 = 25\%$
All forms of $\frac{2}{8}$ have the same value. Changing the form doesn't change its value.	$0.25 = 25\% = \frac{25}{100}$

Remember the Basics

Direct students to the chart at the bottom of the page. In this activity, students fill in the blanks with equivalent fractions, decimals, and percents. If students are having difficulty, review how to convert.

Say, *You need to know how to convert percents, fractions, and decimals in order to solve percent problems. What does* equivalent *mean?* (**equal**)

	FRACTION	DECIMAL	PERCENT
1.	$\frac{1}{1}$	1.00	100%
2.	$\frac{3}{4}$	0.75	**75%**
3.	$\frac{2}{3}$	**0.66**	66%
4.	$\frac{17}{100}$	0.17	**17%**
5.	$\frac{1}{5}$	**0.20**	20%
6.	$\frac{7}{8}$	0.875	**87.5%**

Find the Part Student pages 44 45

LESSON 13

44 Students learn how to fill in a blank percent triangle with the correct arrangement of words and operations.

TO THE CLASS	ON THE BOARD
Every percent problem has 3 pieces of information that you need to know: the part, the whole, and the percent. You can use a percent triangle to find a missing piece of information.	
Cover what is missing and look at the operation sign between the other two pieces.	Cover the word part.
If you're looking for the part, then what two pieces of information do you have?	Point to whole and percent.
What should you do to find the part?	Multiply the whole times the percent to get the part. Point to the × in the triangle.
Is there another way to write this sentence?	whole × percent = part

EXAMPLE Find 25% of 75. (**18.75**)

ASK	ANSWER
Have you seen problems like this on tests? How do you know what pieces of information you have?	The number followed by a % is the percent and the number after *of* is the whole.
What piece of the triangle should you cover?	Part, because that is missing.
What's the percent sentence?	part = whole × percent
What operation will you use to solve the problem?	(Point to the × on the triangle.) Multiplication.
What is 25% of 75?	18.75

On Your Own

There are 60 students in Mr. Watt's math class. 65% of them have jobs after school. How many students have jobs after school? (**39**)

- Direct students to *65% of them* and ask, *Is* them *the part or the whole?* (**whole**) *What does* them *stand for?* (**the students in Mr. Watt's class**)

- Ask, *What piece of information is missing? Write a percent sentence, substitute numbers, and then solve.*

- Direct students to **Practice Find the Part** on page 45. Have them work in groups and correct each other's answers.

LEARNING STYLES

VISUAL Have students use 3 different colored markers in the percent triangle to highlight %, whole, and part. Then have them use the same colors to highlight the corresponding words or numbers in the problem.

Math Toolkit 77

Use the Percent Triangle on student page 77 to help students find the part in a percent problem.

Practice Answers 45

Complete solutions to all Practice problems on student page 45 are at the back of this book on p. 44.

1. 41   2. 20.46
3. 4.5   4. 300
5. 20.7   6. 0.55
7. 3.6   8. 7.764
9. 6   10. 42 seniors
11. 24 actors 12. 25 stores
13. 9 fly balls 14. 98 houses
15. $200,000

LESSON 14

Find the Whole — Student pages 46 47

LEARNING STYLES

VISUAL Have students cut a small square out of colored paper. Tell students to use this square to cover the missing part on the percent triangle when trying to solve problems.

77 Math Toolkit

Use the Percent Triangle on student page 77 to help students find the whole in a percent problem.

47 Practice Answers

Complete solutions to all Practice problems on student page 47 are at the back of this book on p. 44.

1. 84
2. 120
3. 300
4. 12
5. 80
6. 15
7. 96
8. 288
9. 80 students
10. 200 businesses
11. 5,000 applied
12. 200 deer
13. 120 parents
14. 600 students

46 Students learn how to use the percent triangle to find the whole in percent problems.

TO THE CLASS	ON THE BOARD
Sixteen teens have an after-school job at a video store. These 16 teens represent 20% of the students who applied. How many teens applied? Use a percent triangle to find the missing information.	*(Percent triangle: Part ÷ ÷, Whole × %)*
Cover what is missing and look at the operation sign between the other two pieces.	Cover the word whole.
If you're looking for the whole, then what two pieces of information do you have?	Point to part and percent.
What should you do to find the whole? **(Divide the part by the percent.)**	whole = 16 ÷ 20% whole = 16 ÷ 0.20
How many students applied for the job at the video store?	whole = 80 80 students applied.

EXAMPLE *30 is 60% of what number?* **50**

ASK	ANSWER
What pieces do you have?	Part is 30. Percent is 60%.
What percent sentence can you use?	whole = part ÷ percent
What do you get when you replace the words with numbers.	whole = 30 ÷ 60% whole = 30 ÷ 0.60
Find the answer.	whole = 50
30 is 60% of what number?	50

On Your Own

12 is 24% of what number? **(50)**

- Guide students to use the percent triangle and identify the pieces they have and what is missing. **(have part and percent; missing the whole)**
- Guide student to the correct percent sentence. **(whole = part ÷ percent)**
- Tell students to substitute numbers for words and solve the problem. **(whole = 12 ÷ 24% = 12 ÷ 0.24 = 50)**
- Invite a volunteer to solve the percent sentence on the board and explain how he or she arrived at the solution.

Direct students to **Practice Find the Whole** on page 47. Have them work in groups and correct each other's answers.

Find the Percent Student pages 48 49 50

LESSON 15

48 Students learn how to use the percent triangle to find the percent in problems.

TO THE CLASS	ON THE BOARD
200 radio listeners called in to name their favorite group. 40 named the same group as their favorite. What percent of those that called in named this group? You can use a percent triangle to find the missing piece of information.	*(percent triangle: Part ÷ ÷ Whole × %)*
If you are looking for the percent, what two pieces of information do you have? **(the part and the whole)**	Cover the percent symbol.
What should you do to find the percent? **(Divide the part by the whole.)**	Percent = part ÷ whole
How can you find the number of radio listeners who had the same favorite group?	Percent = 40 ÷ 200 Percent = 0.2
What is this answer as a percent?	0.2 × 100 = 20%

EXAMPLE *Thirty-three is what percent of 150?* **(22%)**

TO THE CLASS	ON THE BOARD
What pieces of information do you have?	Part is 33. Whole is 150.
Write this as a percent sentence?	Percent = (part ÷ whole) × 100
What sentence can you write using the information that you have?	Percent = (33 ÷ 150) × 100
Solve the equation to find the percent?	Percent = 0.22 × 100 = 22%
Thirty-three is what percent of 150?	22%

On Your Own

What percent of 88 is 22? **(25%)**

- Say, *Use the percent triangle to identify the pieces that you have.* **(the part and the whole)** Ask, *What part is missing?* **(the percent)**

- Direct students to write the percent sentence needed to solve the problem and to find the missing percent. (**Percent = (part ÷ whole) × 100 = (22 ÷ 88) × 100 = 0.25 × 100 = 25%**)

Direct students to **Practice Find the Percent** on page 49. Have them work in groups and correct each other's answers.

LEARNING STYLES

KINESTHETIC Have students make a model of the percent triangle. To set up percent sentences, have students write the given information on glue-backed note paper. Place the information in the appropriate area of the triangle. Note paper can be replaced with each new problem.

Practice Answers 49

Complete solutions to all Practice problems on student page 49 are at the back of this book on p. 44.

1. 25%   2. 37.5%
3. 25%   4. 70%
5. 5%   6. 16.25%
7. 54%   8. 15%
9. 160%   10. 16%
11. 33.3% 12. 75%
13. 450% 14. 25%

Test-Taking Strategy 50

DRAW A DIAGRAM
You can draw and label a percent triangle to answer questions about percents.

EXAMPLE
Ask for volunteers to read aloud the example problem and the steps.

TRY IT OUT
Have students work out the problem on their own using the percent diagram to help them identify the part. **Answer: D**

LESSON 15 25

LESSONS 16–17 OVERVIEW

Skills

Identify a percent equation.

Write a percent equation from a word problem.

KEY WORDS

equation: a statement of equality between two terms

percent equation: an equation used to show and solve percent problems

variable: a letter used to represent the number you are trying to find.

LEARNING TIPS

- Remember that frequently *of* means *multiply*, and *is* means *equals* in word problems.
- Equations can be written in different forms: $20 \times n = 14$, $n = 14 \div 20$, and $20 = 14 \div n$ are all equations where n is a variable that somehow relates 20 and 14.

English Language Learners

- The word *part* has multiple meanings. Be sure students understand the meaning of the word in the context of percents.
- To avoid confusion with the use of different letters as variables, explain that a variable is usually a meaningful letter. For example, n stands for *number*.
- Make sure that students do not confuse the multiplication sign with the variable for an unknown number, x. Also do not use x as a variable in equations with multiplication.

Percents and Problem-Solving II

Student page 51

Access Prior Knowledge

What do you do to the equation $4n = 20$ to find the value of n? **Divide both sides of the equation by 4; $n = 5$**

Guided Instruction

Direct students to the review of working with percents.

TO THE CLASS	ON THE BOARD
What three parts do you need to solve percent problems?	part, whole, percent
What sentence do you use to find the part?	part = whole × percent
What sentence do you use to find the whole?	whole = part ÷ percent
What sentence do you use to find the percent?	percent = (part ÷ whole) × 100
When given two of the three numbers in a percent problem, you can find the value of the third number, n, by using one of these sentences.	n = whole × percent n = part ÷ percent n = part ÷ whole

Remember the Basics

Direct students to the problems in the chart at the bottom of the page. In this activity, students solve equations for n. If students are having difficulty, review how to use opposite operations to solve equations.

1. $25n = 400$ $\dfrac{25n}{25} = \dfrac{400}{25}$ $n = 16$	2. $n = 0.2 - 120$ $n = -119.8$	3. $48 = 0.6n$ $\dfrac{48}{0.6} = \dfrac{0.6n}{0.6}$ $80 = n$
4. $\dfrac{n}{1.2} = 30$ $1.2 \, \dfrac{n}{1.2} = 30 \times 1.2$ $n = 36$	5. $3.6 = 0.9n$ $\dfrac{3.6}{0.9} = \dfrac{0.9n}{0.9}$ $4 = n$	6. $2.4 + 4 = 0.8n$ $6.4 = 0.8n$ $\dfrac{6.4}{0.8} = \dfrac{0.8n}{0.8}$ $8 = n$

What Is a Percent Equation? Student pages 52 53

LESSON 16

52 Students learn how to write a percent equation.

TO THE CLASS	ON THE BOARD
When the problem says *percent of*, what operation do you use? (multiplication)	25 is what percent of 34? 25 is what decimal multiplied by 34?
When the problem says *is*, what symbol do you use? (Use *equals* for the word *is*.)	25 is what decimal multiplied by 34?
What can you use if the number is unknown? (Use a variable, such as *n*.)	Let n = the unknown number.
Write the percent equation.	$25 = n \times 34$

EXAMPLE *30 whales were spotted offshore. This is 60% of the total expected. How many whales were expected?* (50)

ASK	ANSWER
State the problem in terms of a percent sentence.	30 is 60% of what number?
What does the word *is* mean?	equals
What does the word *of* mean?	multiplied by
What represents the missing number?	n
What is the equation for the problem?	$30 = 60\% \times n$
What do you have to do to the percent before you can solve the problem?	Change the percent to a decimal, then solve. $30 = 0.60 \times n$ $\dfrac{30}{0.60} = \dfrac{0.60n}{0.60}$ $50 = n$
How many whales were expected?	50

On Your Own

Last year Ms. Salas had 30 students in her class. There were 120 students in the school. What percent of the students were is Ms. Salas' class? (25%)

- Direct students to restate the problem as a percent equation. ($n \times 120 = 30$)
- Tell students to solve for the missing number, *n*. ($0.25 = 25\%$)
- Invite a volunteer to solve the problem on the board and explain how he or she arrived at the solution.

Direct students to **Practice What Is a Percent Equation?** on page 53. Have them work in groups and correct each other's answers.

LEARNING STYLES

KINESTHETIC Have students make note cards with *percent, part, whole, n, =,* and × written on them. Have students arrange the cards to create all forms of a percent equation. When there is a missing number, have students use the *n* card in the equation.

Practice Answers 53

Complete solutions to all Practice problems on student page 53 are at the back of this book on p. 44.

1. $n \times 60 = 24$
2. $n = 0.32 \times 80$
3. $n \times 64 = 12$
4. $0.80 \times n = 16$
5. $n \times 240 = 180$
6. $n = 0.28 \times 54$
7. $0.62 \times n = 16$
8. $n \times 20 = 60$
9. $n = 0.02 \times 10$
10. $0.10 \times n = 12$
11. $n \times 3.6 = 1.8$
12. $n = 0.022 \times 200$
13. $0.44 \times n = 22$
14. $n \times 68 = 16$
15. $n = 0.0232 \times 50$
16. $0.75 \times n = 33$

LESSON 17

Creating a Percent Equation from a Word Problem
Student pages 54 55

54 Students learn how to write a percent equation from a word problem.

LEARNING STYLES

AUDITORY Have students work in pairs to do several of the practice problems. One partner should say *part, whole,* and *percent*. The other partner should identify each of these terms for that particular problem. Partners should alternate roles from problem to problem.

TO THE CLASS	ON THE BOARD
When solving a word problem that uses percents, first ask, *What is the problem asking me to find?*	The part? The whole? Or the percent?
What can you look for in the problem?	numbers, percents, words like of or is
What can you do with the information?	Write a percent equation.
What can you use if information is missing?	a variable, such as n

EXAMPLE Maureen has a collection of 80 DVDs. She loaned 16 of them to her friends. What percent of the DVDs did she loan? (**20%**)

ASK	ANSWER
What is the problem asking you to find?	the percent
What numbers are given to you in the problem?	80 and 16
What number is the whole?	80
What number is the part?	16
What percent equation can you write?	$n \times 80 = 16$
What do you get when you solve for n?	$\frac{80n}{80} = \frac{16}{80}$; $n = 0.20$
What percent of the DVDs did Maureen loan?	20%

55 Practice Answers

Complete solutions to all Practice problems on student page 55 are at the back of this book on p. 45.

1. $44 = 0.2 \times n$; 220
2. $0.30 \times 1,340 = n$; 402
3. $n \times 50 = 22$; 44%
4. $0.11 \times n = 55$; 500
5. $0.85 \times 8 = n$; 6.8
6. $n \times 50 = 125$; 2.5% increase
7. $0.002 \times n = 18$; 9,000
8. $n = 0.28 \times 145$; 40.6 = 40
9. $n \times 623 = 386$; 62%
10. $0.945 \times n = 821$; 868.8 = 868

On Your Own

In a parade, 1,170 marchers each carry musical instruments. This is 65% of the total marchers. How many marchers are in the parade? (**1,800**)

- Direct students to identify the information in the problem that they can use to solve it. (**65%; 1,170 marchers**)
- Ask, *What is the missing part?* (**the whole**)
- Tell students to write the percent equation. (**$0.65n = 1,170$**)
- Guide students to find the answer. (**$n = 1,800$**)
- Invite a volunteer to solve the problem on the board and explain how he or she arrived at the solution.

Direct students to **Practice Creating a Percent Equation from a Word Problem** on page 55. Have them work in groups and correct each other's answers.

56 Test-Taking Strategy

CHECK FOR REASONABLENESS OF AN ANSWER
You can check to see that your answers are reasonable when solving percent problems.

EXAMPLE
Ask for volunteers to read aloud the example and the steps.

TRY IT OUT
Have students work out the problem on their own using estimation. Review how to find the right answer. **Answer: C**

Review Test-Taking Strategies

Student pages 36 50 56

Ask for volunteers to name a Test-Taking Strategy they learned in this unit. Have them explain how to use the strategy and why they find it helpful.

USE A GRAPH — 36

A graph can be useful in solving percent problems.

MAKE A DRAWING — 50

Drawing a percent triangle will help when answering questions about percents.

CHECK FOR REASONABLENESS OF AN ANSWER — 56

Checking for reasonableness of an answer helps determine if a problem was solved correctly.

> **USE A STRATEGY**
> - Divide students into small groups and assign them one of the strategies in the unit.
> - Using the model from their books, have each group create 3 new problems. Have the group solve each problem on a separate sheet of paper.
> - Invite groups to exchange and solve problems.
> - After groups have finished solving problems, they can exchange their solutions.

Mixed-Level Classroom

Student pages 29 37 43 51

REVISIT LEARNING OBJECTIVES

Review the learning objectives in Unit 2. You may want to refer students to the Overview pages in this Unit.

OVERVIEW	LEARNING OBJECTIVES	SE PAGE
Lessons 8–10	• Use percent and fractions show a ratio out of 100. • Use proportions to write fractions as percents. • Write percents as fractions.	29
Lessons 11–12	• Write any decimal as a percent. • Write any percent as a decimal.	37
Lessons 13–15	• Find the whole, the part, and the percent. • Apply strategies to help solve percent problems.	43
Lessons 16–17	• Identify a percent equation. • Write a percent equation from a word problem.	51

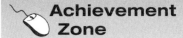

Achievement Zone

You can download reproducible worksheets that reinforce converting fractions, decimals, and percents at **www.HarcourtAchieve.com/AchievementZone**.

RETEACH

Write these percent problems on the board and invite students to solve them using the percent triangle.

1. What is 10% of 600? **(60)**
2. What is 12.5% of 56? **(7)**
3. Find 200% of 16. **(32)**
4. Find 10% of $98. **($9.80)**

UNIT REVIEW ANSWERS — 57

1. 47%   2. 88%   3. 65%   4. 70%   5. $\frac{9}{25}$ 6. 42%   7. 0.61
8. 0.048   9. 168   10. 160%   11. 45   12. 100%   13. 32 CDs
14. $216   15. $n \times 18 = 8; n = 44.4\%$ 16. $0.20 \times n = 6; n = 30$ hours

> **REMIND STUDENTS**
> - You have to use the decimal form of the percent in operations.
> - You multiply the whole by the decimal form of the percent.
> - You divide the part by the decimal form of the percent.

UNIT 3

Percents in Daily Life

Student page 58

Students recognize the use of percents in real-life situations by considering how they deal with percents as consumers. *How often do you come across percents in your daily life?*

Learning Objectives

In Unit 3, students will
- Determine percent change.
- Calculate markups and discounts.
- Determine simple interest.
- Compute compound interest.

Real-Life Matters

Invite volunteers to read Real-Life Matters on page 58. Ask students how to use percents to determine discounts.

ASK	ANSWER
What does the sign in the photograph mean? Is the new price more or less than the original price?	It means that you take the original price and take away 50%. The photograph means that the new price is less than the original price.
What is 50% expressed as a decimal? As a fraction?	Decimal = 0.5 Fraction = $\frac{1}{2}$
How would you determine how much is taken off the original price?	Take the original price and multiply it by 0.5. That tells what the discount is.
Suppose you went into the store and saw a pair of jeans that had an original price of $30.00. What would the discount be?	$30 \times 0.5 = 15$ The discount is $15.00.

LEARNING TIPS

- Discounted prices are always less than original prices.
- Marked-up prices are always greater than original prices.
- Interest is calculated in more than one way.

Real-Life Application

Students answer questions that ask them to determine and compare discounted prices. Let students work in small groups to brainstorm answers to questions presented. Invite groups to share their answers with the class. You might launch with the following questions:

- *What must you do to the percent in order to calculate the discount?* (**change the percent to a decimal**)

- *How do you change a percent to a decimal?* (**move the decimal point in the percent two places to the left**)

- *So, what is 20% expressed as a decimal?* (**20% = 0.2**)

- *When you calculate the discount is this the final price?* (**No; it is the amount that is subtracted from the original price.**)

- *So what must you do to find the final price?* (**the original price − discount**)

FOLLOW-UP

Would you save more if you had a discount of 20% on a $42 pair of jeans or if the discount was 15% on a $47 pair of jeans? **($8.40 or $7.05; you save more with the 20% discount.)**

OVERVIEW LESSONS 18–19

Application of Percents I Student page 59

Skills
Find the percent of increase or decrease.
Determine the markup or discount.

Access Prior Knowledge
What form must a percent take to be used in a calculation? **A percent must be written as a decimal. 25% = 0.25**

Guided Instruction
Direct students to the clothing store illustration.

TO THE CLASS	ON THE BOARD
If the original price of a jacket is $125 and the sale price is $100, what is the amount of the discount?	$125 − $100 = $25
To find the percent of discount, do you compare the $25 to the original price or the sale price?	the original price
What equation can be used to calculate the percent of discount?	whole × percent = part 125 × n = 25
How do you solve this equation for *n*?	$\frac{125n}{125} = \frac{25}{125}$ $n = \frac{25}{125} = \frac{1}{5} = 0.20 = 20\%$
What percent is the discount?	20%

KEY WORDS
markup: the difference between what a store pays and the selling price
discount: a cut in price

LEARNING TIPS
- Remember to change the percent to a decimal before solving percent problems.
- Newspapers are good sources for discount examples.
- Comparative shopping requires evaluation of percents of discount.

Remember the Basics
Direct students to the problems at the bottom of the page. In this activity, students solve percent problems involving percents and money amounts. If students are having difficulty, review how to identify the part, the whole, and the percent. Say, *You need to know the whole, part, and percent to solve percent problems. What type of equation is used to solve these problems?* **(percent equation)**

English Language Learners
- Using numbers and pictures show students examples of discounts and markups. Cross out the highest priced items. Have students circle what they believe to be the best price.

1. What is 20% of $120? $n = 0.2 \times 120 = 24$ $n = \$24$ 20% of $120 = **$24**.	**2.** What percent of $96 is $16? $n \times 96 = 16$ $n = 0.167 = $ **16.7%**
3. What percent of $40 is $32? $n \times 40 = 32$ $n = 0.80 = $ **80%**	**4.** $15 is 30% of what dollar amount? $0.30 \times n = 15$ $n = $ **$50**
5. $12 is 40% of what dollar amount? $0.40 \times n = 12$ $n = $ **$30**	**6.** What is 25% of $550? $0.25 \times 550 = n$ $n = $ **137.5 or $137.50**

LESSON 18

Percent Change   Student pages 60 61

LEARNING STYLES

AUDITORY Give the students index cards with percent problems on them. Have pairs of students read aloud to each other the information on the cards. When one student reads, the other student tells whether there is a percent increase or decrease.

60 Students learn how to find percents of increase and decrease.

TO THE CLASS	ON THE BOARD
When finding percent of increase or decrease, look at the original and final numbers. A company started with 30 employees and now has 60 employees.	Original number: 30 Final number: 60
How has the number of employees changed?	increased by $60 - 30 = 30$
Write a sentence to find the percent increase?	30 is what percent of 30?
What is this sentence as an equation?	$30 = n \times 30$
Solve for n. Solve the equation for a fraction, then a decimal, and then a percent.	$\frac{30}{30} = \frac{30n}{30}$ $n = \frac{30}{30} = \frac{1}{1} = 1.00 = 100\%$
What is the percent of increase?	100%

EXAMPLE A company that had 30 workers now has only 24. What is the percent of decrease? (**20%**)

ASK	ANSWER
What information do you know?	30 workers originally 24 workers now
How has the number of workers changed?	decreased by 6
What do you compare to the amount of decrease?	the original number, 30
How could you state this problem as a sentence?	6 is what percent of 30?
What is this sentence as an equation?	$6 = n \times 30$
How do you solve for n?	$n = \frac{6}{30} = \frac{1}{5} = 0.20 = 20\%$
What is the percent of decrease?	20%.

On Your Own

There were 40 stores in a mall. Now there are 35. What is the percent of decrease?

- Direct students to find needed information from the problem. (**40 is original number; 35 is final number.**)

- Ask, *Was there an increase or decrease?* (**40 − 35 = 5; decrease of 5**)

- Tell students to write the appropriate percent equation and solve the problem. (**5 = n × 40; n = 0.125 or 12.5%**)

- Invite a volunteer to solve the problem on the board and explain how he or she arrived at the solution.

Direct students to **Practice Percent Change** on page 61. Have them work in groups and correct each other's answers.

61 Practice Answers

Complete solutions to all Practice problems on student page 61 are at the back of this book on p. 45.

1. Decrease is 5%.
2. Decrease is 25%.
3. Increase is 6.25%.
4. Decrease is 50%.
5. Increase is 20%.
6. Decrease is 16%.
7. Decrease is 16.7%.
8. Decrease is 10%.
9. Decrease is 10%.
10. Increase is 25%.
11. Increase is 10%.
12. Decrease is 26.7%.
13. Decrease is 10%.
14. Decrease is 20%.

Markup and Discount

LESSON 19

Student pages 62 63

62 Students learn how to compute markups and discounts.

ASK	ANSWER
What is another name for percent markup?	% increase
What steps would you take to find the percent markup?	1. Find the amount of markup. 2. Write an equation using the cost to the store, the dollar amount of markup, and the percent markup. percent × cost = markup 3. Solve the equation.
What is another name for percent discount?	% decrease
What changes in the steps for finding percent markup would you make to find percent discount?	Compare the amount of decrease to the selling price. percent × cost = decrease

EXAMPLE *A store receives cell phones for $24 each and sells them for $36 each. What is the percent of markup?* **(50%)**

TO THE CLASS	ON THE BOARD
What is the amount of markup (increase)?	$36 − $24 = $12
What you are trying to find?	What percent of 24 is 12?
How would you write this sentence as an equation?	n × 24 = 12
How would you solve this equation for *n*?	$\frac{24n}{24} = \frac{12}{24}$
Your answer is in the form of a fraction. How would you change it to a percent?	$n = \frac{1}{2} = 0.50 = 50\%$
What is the percent of markup on the cell phones?	50%

On Your Own

An $800 stereo system now sells for $750. What is the percent of discount? **(6.25%)**

- Direct students to find the amount of decrease. ($800 − $750 = $50) Use the equation to find the value of *n*, the percent of discount.
- Ask, *What percent equation should you use?* ($50 = n × $800)
- Tell students to solve for the percent. ($n = \frac{50}{800} = 0.0625 = 6.25\%$)
- Ask, *What is the percent of discount?* **(6.25%)**
- Invite a volunteer to solve the problem on the board and explain how he or she arrived at the solution.

Direct students to **Practice Markup and Discount** on page 63. Have them work in groups and correct each other's answers.

LEARNING STYLES

VISUAL Have the students make a markup/discount chart. Tell them to write *PRICE*. From that word, have them draw an arrow pointing down. By the arrow, write *DISCOUNT*. Then have them draw an arrow pointing up, and label it *MARKUP*.

Practice Answers 63

Complete solutions to all Practice problems on student page 63 are at the back of this book on p. 46.

1. The decrease is 15.8%.
2. The decrease is 20%.
3. The decrease is 10%.
4. The increase is 25%.
5. The increase is 30%.
6. The decrease is 30%.
7. The increase is 80%.
8. The increase is 16.7%.
9. The discount is 20%.
10. The markup is 20%.
11. The discount is 20%.
12. The markup is 40%.
13. The discount is 14.6%.
14. The discount is 22.2%.

Test-Taking Strategy 64

CHOOSE A STRATEGY
Solve percent problems using a strategy that works best for you.

EXAMPLE
Ask for volunteers to read the steps aloud.

TRY IT OUT
Have students work out the problem on their own using a strategy of their choice. Review how to find the right answer.
Answer: C

LESSON 19 33

LESSONS 20–21 OVERVIEW

Skills

Find simple interest and compound interest

Use interest formulas.

KEY WORDS

interest: money paid for using or saving other money

simple interest: money paid based only on the principal

compound interest: money paid based on the principal plus the interest already earned

principal: the amount of money you start with

rate: the percent of interest paid on the principal

time: the amount of time money is in the bank

balance: the principal plus any interest earned

LEARNING TIPS

- When working with money amounts, change the money to a decimal.
- Change percents to decimals before performing operations.
- Remember the formula:
  interest = principal × rate × time

English Language Learners

- The word *interest* and *principal* have several different meanings. Be sure students understand these words and their meanings in these lessons.
- Have English language learners pair with English speakers who can help them with the terminology associated with percents and interest.

Applications of Percents II Student page 65

Access Prior Knowledge

What is a rate? **A rate is a ratio that compares two numbers with different units: 30 dollars per month.**

Guided Instruction

Direct students to the interest example.

TO THE CLASS	ON THE BOARD
When do banks pay customers?	when they have a savings account
When do banks charge customers?	when they take out loans
What is it called when you have a savings account and you earn money on only the amount you put in the bank?	simple interest
What is it called when you earn money on both the amount you put in the bank and the amount the bank pays you?	compound interest
Suppose you put money in a savings account and left it there for five years. Would you have more after five years with simple interest or compound interest?	compound interest

Remember the Basics

Direct students to the table at the bottom of the page. In this activity, students solve problems using percents and money amounts. If students are having difficulty, review how to write and solve percent equations.

It is important that you know about percents when trying to save money. What does the word interest *mean when referring to percents and money?* (**It is the amount of money paid or charged based on the original amount of money and the interest rate which is given as a percent.**)

1. What is 22% of $400? $22\% = 0.22$ $n = 0.22 \times 400$ $n = 88$ 22% of $400 is $88.	2. Find 4% of $300. $4\% = 0.04$ $n = 0.04 \times 300$ $n = 12$ 4% of $300 is $12.	3. Find 2.5% of $600. $2.5\% = 0.025$ $n = 0.025 \times 600$ $n = 15$ 2.5% of $600 is $15.
4. $0.04 \times 2{,}500 \times 3 =$ **300**	5. $\frac{1}{4} \times 200 \times 2 =$ **100**	6. $0.055 \times 1500 \times 4 =$ **330**

34 OVERVIEW LESSONS 20–21

Simple Interest
Student pages 66 67

LESSON 20

66 Students learn how to compute simple interest.

TO THE CLASS	ON THE BOARD
You put $200 in the bank. What is the name for the money you invest?	$200 = principal (p)
The bank pays you 4% on your $200. What is this 4% called?	4% = rate (r)
If you keep the money in the bank for 3 years, what is t?	3 years = time (t)
The bank pays you money based on these three quantities. What is this money called?	interest (I)
Use this formula to find the interest earned.	$I = p \times r \times t$
When you multiply by a percent, write it as a decimal.	$I = \$200 \times 0.04 \times 3 = \$24$

EXAMPLE *How much simple interest will you earn when you put $600 in your bank at 4% interest for 6 months?* (**$12**)
How much money will be in this account at the end of 6 months? (**$612**)

ASK	ANSWER
What is the principal?	$600
What is the rate? Write it as a decimal.	4% = 0.04
How long in years is the principal invested?	6 months = 0.5 years
What equation do you use to solve for interest?	$I = p \times r \times t$
Solve the equation. What is the interest?	$I = \$600 \times 0.04 \times 0.5 = \$12$
How do you find the total amount in the account at the end of 6 months?	Add the principal and the interest. $600 + $12 = $612

On Your Own

You save $3,000 for 2 years in a bank that has an interest rate of 5% per year. Find the interest. (**$300**) Find the total amount in your account after two years. (**$3,300**)

- Direct students to find the needed information in the problem (**$3,000 is the principal, 5% is the rate, 2 years is the time**) and to use the interest formula to find the interest ($I = p \times r \times t$; $I = 3,000 \times 0.05 \times 2 = 300$).

- Ask, *What is the total amount after 2 years?* (**$3,000 + $300 = $3,300**)

Direct students to **Practice Simple Interest** on page 67. Have them work in groups and correct each other's answers.

LEARNING STYLES

VISUAL Have groups of students write different principal amounts, interest rates, and times on index cards. Ask students to keep the cards in groups, face down. Have them draw a principal, an interest rate, and a time. They can arrange the cards according to the formula for simple interest and calculate the interest amount.

Practice Answers 67

Complete solutions to all Practice problems on student page 63 are at the back of this book on p. 47.

1. $48
2. $160
3. $50
4. $12.25
5. $160; $1,160
6. $337.50; $1,237.50
7. $70.20; $1,370.20
8. $54; $854
9. $377.60
10. $22.50
11. $74.25; $1,174.25
12. $21.60
13. $1.69
14. $1,168.75

LESSON 20 35

LESSON 21

Compound Interest
Student pages 68 69

68 Students learn how to find an account balance when interest is compounded.

LEARNING STYLES

KINESTHETIC Divide the class into two groups, the simple interest group and the compound interest group. Give each group 200 beans. Tell them that the interest on these beans will be calculated yearly at a rate of 4% for 5 years. Have students keep a chart of the balance along with adding the interest beans after each calculation. **(The simple interest group gains 40 beans. The compound interest group gains 43 beans.)**

ASK	ANSWER
Compound interest differs from simple interest. How is it different?	Compound interest is paid on the principal plus earned interest.
What is the compound interest formula?	$p(1 + r)^t = B$
What should you compute first?	Compute what is in parentheses first. Find $1 + r$.
What should you compute next?	Compute using the exponent. Find $(1 + r)^t$.
Finally, to find the balance, what should you multiply this number by?	the principal Find $p(1 + r)^t = B$.

EXAMPLE You put $3,000 in a savings account that pays 4% interest compounded annually. What is your balance after three years? (**$3,374.59**)

TO THE CLASS	ON THE BOARD
What is the principal?	$p = \$3,000$
What is the rate? What is the rate as a decimal?	$r = 4\% = 0.04$
How often is the interest compounded?	once a year
How many years is the money invested?	$t = 3$; 3 years
What is the equation for compound interest using these values?	$B = p(1 + r)^t$ $B = 3,000(1 + 0.04)^3$
Using the order of operations, solve for B.	$B = 3,000(1.04)^3$ $= 3,374.59$
What is your balance after three years?	$3,374.59

69 Practice Answers

Complete solutions to all Practice problems on student page 69 are at the back of this book on p. 47.

1. $648.96
2. $1,071.90
3. $1,095.03
4. $2,272.50
5. $2,090
6. $4,943.40
7. $6,083.50
8. $5,365.63
9. $1,586.08
10. $338.04
11. $1,153.70
12. $10,129.50
13. $131,501.20
14. $159.26

70 Test-Taking Strategy

MAKE A TABLE
You can make a table to help you answer questions about interest.

EXAMPLE
Ask for volunteers to read the example and the steps aloud.

TRY IT OUT
Have students work out the problem on their own using a table. Review how to find the right answer. **Answer: C**

On Your Own

You put $400 in the bank for 4 years at a rate of 5% compounded annually. Find the total amount in your account after four years. (**$486.20**)

- Direct students to find the needed information from the problem. (**$400 is the principal, 5% is the rate, 4 years is the time.**)
- Tell students to use the compound interest formula to find the balance. ($B = p(1 + r)^t$: $B = 400(1 + 0.05)^4 = 486.20$ or **$486.20**)
- Invite a volunteer to solve the problem on the board and explain how he or she arrived at the solution.

Direct students to **Practice Compound Interest** on page 69. Have them work in groups and correct each other's answers.

Review Test-Taking Strategies

Student pages 64 70

Ask for volunteers to name a Test-Taking Strategy they learned in this unit. Have them explain how to use the strategy and why they find it helpful.

CHOOSE A STRATEGY   64

Choose the best strategy that works for you to solve percent problems.

MAKE A TABLE   70

You can make a table to help answer questions about interest.

USE A STRATEGY

- Poll students for their favorite strategy
- Divide students into small groups and assign them one of the strategies
- have each student group create a set of problems for another group to solve using one of the strategies
- Ask for volunteers to share their strategic problem solving on the board.

Mixed-Level Classroom

Student pages 59 65

REVISIT LEARNING OBJECTIVES

You may want to refer students to the Overview pages in this unit. For further reflection, have students complete *Review Your Progress* on the inside back cover of their student book, *Ratios and Percents*.

OVERVIEW	LEARNING OBJECTIVES	PAGE
Lessons 18–19	• Find the percent of increase or decrease. • Determine the markup or discount.	59
Lessons 20–21	• Find simple interest and compound interest.	65

RETEACH

Write these problems on the board and invite students to solve them. Have them compare the final balances for simple interest and compound interest.

Find simple interest and the total amount.

1. $500 is invested for 2 years at 4.5%. (**$45, $545**)
2. $3,000 is invested in a college fund for 18 years at 5%. (**$2,700; $5,700**)

Find the balance.

3. $500 is invested for 2 years at 4.5% compounded annually. (**$546.01**)
4. $3,000 is invested for 18 years at 5% compounded anually. (**$7,297.61**)

UNIT REVIEW ANSWERS   71

1. The decrease is 50%.   2. The increase is 100%.
3. The markup is 37.5%.   4. The markup is 37.5%.
5. The discount is 20%.   6. The discount is 15%.   7. $54   8. $36
9. $12; $212   10. $337.50; $2,837.50   11. $252.50   12. $578.80
13. The increase is 6.25%.   14. The discount is 26.7%.   15. $575
16. $675.30

Achievement Zone

You can download reproducible worksheets that reinforce working with simple and compound interest at

www.HarcourtAchieve.com/ AchievementZone.

REMIND STUDENTS

- Use the decimal form of the percent in operations.
- Always use the correct order of operations to solve problems.

ANSWERS AND EXPLANATIONS

PRETEST

PAGES 4–5

1. $\frac{3}{4}$, 3:4, 3 to 4   2. $\frac{4}{12}$, 4:12, 4 to 12 or $\frac{1}{3}$, 1:3, 1 to 3
3. $\frac{12 \text{ pages}}{3 \text{ minutes}} = \frac{12 \text{ pages} \div 3}{3 \text{ minutes} \div 3} = \frac{4 \text{ pages}}{1 \text{ minute}}$
4. $\frac{80 \text{ kilometers}}{4 \text{ hours}} = \frac{80 \text{ kilometers} \div 4}{4 \text{ hours} \div 4} = \frac{20 \text{ kilometers}}{1 \text{ hour}}$
5. $\frac{\$1.32}{6 \text{ cans}} = \frac{\$1.32 \div 6}{6 \text{ cans} \div 6} = \frac{\$0.22}{\text{can}}$
   The rate is 22 cents per can.
6. $\frac{\$2.10}{3 \text{ pounds}} = \frac{\$2.10 \div 3}{3 \text{ pounds} \div 3} = \frac{\$0.70}{\text{pound}}$
   The rate is 70 cents per pound.
7. $3 \times 10 = 30; 6 \times 8 = 48$
   $\frac{3}{8} \neq \frac{6}{10}$
8. $4 \times 27 = 108; 9 \times 12 = 108$
   $\frac{4}{9} = \frac{12}{27}$
9. $9 \times n = 6 \times 6$
   $9n = 36$
   $\frac{9n}{9} = \frac{36}{9}$
   $n = 4$
10. $15 \times n = 20 \times 45$
    $15n = 900$
    $\frac{15n}{15} = \frac{900}{15}$
    $n = 60$
11. $\frac{2}{9} = \frac{n}{225}$
    $9 \times n = 2 \times 225$
    $9n = 450$
    $\frac{9n}{9} = \frac{450}{9}$
    $n = 50$ students
12. $\frac{420}{35} = \frac{n}{1}$
    $35 \times n = 1 \times 420$
    $35n = 420$
    $\frac{35n}{35} = \frac{420}{35}$
    $n = \$12$ per hour
13. $0.065 = 6.5\%$
14. $\frac{7}{8} = 7 \div 8 = 0.875 = 87.5\%$
15. $32\% = \frac{32}{100} = \frac{32 \div 4}{100 \div 4} = \frac{8}{25}$
16. $125\% = \frac{125}{100} = \frac{125 \div 25}{100 \div 25} = \frac{5}{4}$
17. $22\% \times 60 = 0.22 \times 60 = 13.2$
18. $(24 \div 40) \times 100\% = 0.60 \times 100\% = 60\%$
19. $1.2 \div 40\% = 1.2 \div 0.40 = 3$
20. $30 \div 75\% = 30 \div 0.75 = 40$ DVDs
21. $100\% - 15\% = 85\%$
    $40 \times 85\% = 40 \times 0.85 = \$34$
22. $I = \$600 \times 0.04 \times 2 = \$48$
23. $I = \$4,000 \times 0.055 \times 0.75 = \$165$
24. $B = 250(1 + 0.025)^2 = 250(1.025)^2 = 250(1.0506) = \$262.65$

LESSON 1

PAGE 9

1. There are 2 juniors and 5 freshmen. So the ratio of juniors to freshmen is 2 : 5, 2 to 5, or $\frac{2}{5}$.
2. There are 2 juniors and 5 seniors. So the ratio of juniors to seniors is 2 : 5, 2 to 5, or $\frac{2}{5}$.
3. There are 6 sophomores and 5 seniors. So the ratio of sophomores to seniors is 6 : 5, 6 to 5, or $\frac{6}{5}$.
4. There are 6 sophomores and 18 riders. So the ratio of sophomores and riders is $\frac{6}{18}$. Simplified, the ratio is $\frac{6}{18} \div \frac{6}{6} = \frac{1}{3}$, 1 : 3, or 1 to 3.
5. $\frac{5}{3}$, 5 : 3, 5 to 3
6. $\frac{5}{6}$, 5 : 6, 5 to 6
7. $\frac{3}{6}$, 3 : 6, 3 to 6 or $\frac{1}{2}$, 1 : 2, 1 to 2
8. $\frac{6}{8}$, 6 : 8, 6 to 8 or $\frac{3}{4}$, 3 : 4, 3 to 4
9. The ratio of letters is 7 : 11, 7 to 11, or $\frac{7}{11}$.
10. The ratio of Kevin's laps to Sarah's laps is $\frac{25}{35}$. Simplified, the ratio is $\frac{25}{35} \div \frac{5}{5} = \frac{5}{7}$, 5 : 7, or 5 to 7.
11. The ratio of Ethan's points to Ben's points is $\frac{8}{12}$. Simplified, the ratio is $\frac{8}{12} \div \frac{4}{4} = \frac{2}{3}$, 2 : 3, or 2 to 3.
12. The ratio of winning costumes to people on the team is $\frac{7}{11}$, 7 : 11, or 7 to 11.

LESSON 2

PAGE 11

1. $\frac{10 \text{ miles}}{6 \text{ hours}} \div \frac{2}{2} = \frac{5 \text{ miles}}{3 \text{ hours}}$
   The rate is 5 miles in 3 hours.
2. $\frac{20 \text{ dollars}}{6 \text{ books}} = \frac{20 \text{ dollars} \div 2}{6 \text{ books} \div 2} = \frac{10 \text{ dollars}}{3 \text{ books}}$
   The rate is \$10 for every 3 books.
3. $\frac{9 \text{ free throws}}{24 \text{ attempts}} = \frac{9 \text{ free throws} \div 3}{24 \text{ attempts} \div 3} = \frac{3 \text{ free throws}}{8 \text{ attempts}}$
   The rate is 3 free throws for every 8 attempts.
4. $\frac{50 \text{ liters}}{4 \text{ minutes}} = \frac{50 \text{ liters} \div 2}{4 \text{ minutes} \div 2} = \frac{25 \text{ liters}}{2 \text{ minutes}}$
   The rate is 25 liters for every 2 minutes.

5. $\frac{\$32}{4 \text{ hours}} = \frac{\$32 \div 4}{4 \text{ hours} \div 4} = \frac{\$8}{1 \text{ hour}}$
   The rate is $8 in an hour.

6. $\frac{5 \text{ tickets}}{\$80} = \frac{5 \text{ tickets} \div 5}{\$80 \div 5} = \frac{1 \text{ ticket}}{\$16}$
   The rate is 1 ticket for $16.

7. $\frac{6 \text{ cups flour}}{4 \text{ eggs}} = \frac{6 \text{ cups flour} \div 2}{4 \text{ eggs} \div 2} = \frac{3 \text{ cups flour}}{2 \text{ eggs}}$
   The rate is 3 cups of flour for 2 eggs.

8. $\frac{\$300}{4 \text{ tires}} = \frac{\$300 \div 4}{4 \text{ tires} \div 4} = \frac{\$75}{1 \text{ tire}}$
   The rate is $75 for each tire.

9. $\frac{10 \text{ laps}}{5 \text{ minutes}} = \frac{10 \text{ laps} \div 5}{5 \text{ minutes} \div 5} = \frac{2 \text{ laps}}{1 \text{ minute}}$
   The rate is 2 laps in 1 minute.

10. $\frac{80 \text{ miles}}{2 \text{ hours}} = \frac{80 \text{ miles} \div 2}{2 \text{ hours} \div 2} = \frac{40 \text{ miles}}{1 \text{ hour}}$

11. $\frac{10 \text{ rolls}}{4 \text{ hours}} = \frac{10 \text{ rolls} \div 2}{4 \text{ hours} \div 2} = \frac{5 \text{ rolls}}{2 \text{ hours}}$

12. $\frac{15 \text{ laps}}{20 \text{ minutes}} = \frac{15 \text{ laps} \div 5}{20 \text{ minutes} \div 5} = \frac{3 \text{ laps}}{4 \text{ minutes}}$

13. $\frac{\$105}{14 \text{ days}} = \frac{\$105 \div 7}{14 \text{ days} \div 7} = \frac{\$15}{2 \text{ days}}$

14. $\frac{30 \text{ squirrels}}{4 \text{ hours}} = \frac{30 \text{ squirrels} \div 2}{4 \text{ hours} \div 2} = \frac{15 \text{ squirrels}}{2 \text{ hours}}$

15. $\frac{30 \text{ gallons}}{510 \text{ miles}} = \frac{30 \text{ gallons} \div 30}{510 \text{ miles} \div 30} = \frac{1 \text{ gallon}}{17 \text{ miles}}$

8. $\frac{144 \text{ miles}}{6 \text{ gallons}} = \frac{144 \text{ miles} \div 6}{6 \text{ gallons} \div 6} = \frac{24 \text{ miles}}{\text{gallon}}$
   The rate is 24 miles per gallon.

9. $\frac{90 \text{ pages}}{180 \text{ minutes}} = \frac{90 \text{ pages} \div 180}{180 \text{ minutes} \div 180} = \frac{0.5 \text{ pages}}{\text{minute}}$
   The rate is 0.5 pages per minute.

10. $\frac{\$4.80}{12 \text{ bottles}} = \frac{\$4.80 \div 12}{12 \text{ bottles} \div 12} = \frac{\$0.40}{\text{bottle}}$
    The rate is 40 cents per bottle.

11. $\frac{60 \text{ pages}}{30 \text{ minutes}} = \frac{60 \text{ pages} \div 30}{30 \text{ minutes} \div 30} = \frac{2 \text{ pages}}{\text{minute}}$
    The rate is 2 pages per minute.

12. $\frac{500 \text{ words}}{10 \text{ minutes}} = \frac{500 \text{ words} \div 10}{10 \text{ minutes} \div 10} = \frac{50 \text{ words}}{\text{minute}}$
    The rate is 50 words per minute.

13. $\frac{\$9}{6 \text{ slices}} = \frac{\$9 \div 6}{6 \text{ slices} \div 6} = \frac{\$1.50}{\text{slice}}$
    The rate is $1.50 a slice.

14. $\frac{\$350}{4 \text{ rooms}} = \frac{\$350 \div 4}{4 \text{ rooms} \div 4} = \frac{\$87.50}{\text{room}}$
    The rate is $87.50 per room.

15. $\frac{\$4}{5 \text{ pounds}} = \frac{\$4 \div 5}{5 \text{ pounds} \div 5} = \frac{\$0.80}{\text{pound}}$
    The rate is 80 cents per pound.

LESSON 3

PAGE 13

1. $\frac{60 \text{ miles}}{4 \text{ hours}} \div \frac{4}{4} = \frac{15 \text{ miles}}{1 \text{ hour}}$
   The rate is 15 miles per hour.

2. $\frac{\$45}{5 \text{ hours}} = \frac{\$45 \div 5}{5 \text{ hours} \div 5} = \frac{\$9}{\text{hour}}$
   The rate is $9 per hour.

3. $\frac{200 \text{ words}}{4 \text{ minutes}} = \frac{200 \text{ words} \div 4}{4 \text{ minutes} \div 4} = \frac{50 \text{ words}}{\text{minute}}$
   The rate is 50 words per minute.

4. $\frac{88 \text{ miles}}{2 \text{ hours}} = \frac{88 \text{ miles} \div 2}{2 \text{ hours} \div 2} = \frac{44 \text{ miles}}{\text{hour}}$
   The rate is 44 miles per hour.

5. $\frac{\$1.80}{6 \text{ cans}} = \frac{\$1.80 \div 6}{6 \text{ cans} \div 6} = \frac{\$0.30}{\text{can}}$
   The rate is 30 cents per can.

6. $\frac{36 \text{ grams}}{6 \text{ servings}} = \frac{36 \text{ grams} \div 6}{6 \text{ servings} \div 6} = \frac{6 \text{ grams}}{\text{serving}}$
   The rate is 6 grams of fat per serving.

7. $\frac{3{,}000 \text{ meters}}{10 \text{ minutes}} = \frac{3{,}000 \text{ meters} \div 10}{10 \text{ minutes} \div 10} = \frac{300 \text{ meters}}{\text{minute}}$
   The rate is 300 meters per minute.

LESSON 4

PAGE 17

1. $2 \times 18 = 36$; $3 \times 12 = 36$
   Because $36 = 36$, the ratios are a proportion.

2. $18 \times 3 = 54$; $6 \times 8 = 48$
   Because $54 \neq 48$, the ratios are not proportional.

3. $2 \times 20 = 40$; $8 \times 5 = 40$
   Because $40 = 40$, the ratios are a proportion.

4. The ratios are $\frac{6}{8}$ and $\frac{12}{16}$.
   $6 \times 16 = 96$; $8 \times 12 = 96$
   Because $96 = 96$, the ratios are a proportion.

5. The ratios are $\frac{4}{9}$ and $\frac{12}{26}$.
   $4 \times 26 = 104$; $9 \times 12 = 108$
   Because $104 \neq 108$, the ratios are not proportional.

6. $3 \times 6 = 18$; $7 \times 14 = 98$
   Because $18 \neq 98$, the ratios are not proportional.

7. $7 \times 6 = 42$; $2 \times 21 = 42$
   Because $42 = 42$, the ratios are a proportion.

8. The ratios are $\frac{9}{5}$ and $\frac{27}{9}$.
   $9 \times 9 = 81$; $5 \times 27 = 135$
   Because $81 \neq 135$, the ratios are not proportional.

9. $2 \times 80 = 160$; $8 \times 20 = 160$
   Because $160 = 160$, the ratios are proportional.

ANSWERS AND EXPLANATIONS 39

10. $40 \times 6 = 240$; $4 \times 60 = 240$
    Because $240 = 240$, the ratios are proportional. They are paid at the same rate.
11. The ratios are $\frac{3}{10}$ and $\frac{5}{20}$.
    $3 \times 20 = 60$; $10 \times 5 = 50$
    Because $60 \neq 50$, the ratios are not proportional. They do not save money at the same rate.
12. The ratios are $\frac{6}{15}$ and $\frac{9}{21}$.
    $6 \times 21 = 126$; $9 \times 15 = 135$
    Because $126 \neq 135$, the ratios are not proportional. They do not complete passes at the same rate.
13. The ratios are $\frac{50}{1}$ and $\frac{200}{4}$.
    $50 \times 4 = 200$; $1 \times 200 = 200$
    Because $200 = 200$, the ratios are proportional. Their typing speeds are the same.
14. The ratios are $\frac{0.5}{10}$ and $\frac{1}{25}$.
    $0.5 \times 25 = 12.5$; $10 \times 1 = 10$
    Because $12.5 \neq 10$, the ratios are not proportional. They do not walk at the same rate.
15. The ratios are $\frac{8}{10}$ and $\frac{15}{20}$.
    $8 \times 20 = 160$; $10 \times 15 = 150$
    Because $160 \neq 150$, the ratios are not proportional. They do not have the same ratio of correct answers.

LESSON 5

PAGE 19

1. $4 \times n = 3 \times 12$
   $4n = 36$
   $\frac{4n}{4} = \frac{36}{4}$
   $n = 9$
   $\frac{3}{4} = \frac{9}{12}$
2. $6 \times n = 4 \times 9$
   $6n = 36$
   $\frac{6n}{6} = \frac{36}{6}$
   $n = 6$
   $\frac{4}{6} = \frac{6}{9}$
3. $16 \times n = 4 \times 8$
   $16n = 32$
   $\frac{16n}{16} = \frac{32}{16}$
   $n = 2$
   $\frac{8}{16} = \frac{2}{4}$
4. $6 \times x = 5 \times 30$
   $6x = 150$
   $\frac{6x}{6} = \frac{150}{6}$
   $x = 25$
   $\frac{25}{30} = \frac{5}{6}$

5. $30 \times n = 18 \times 10$
   $30n = 180$
   $\frac{30n}{30} = \frac{180}{30}$
   $n = 6$
   $\frac{18}{30} = \frac{6}{10}$
6. $8 \times n = 40 \times 5$
   $8n = 200$
   $\frac{8n}{8} = \frac{200}{8}$
   $n = 25$
   $\frac{40}{25} = \frac{8}{5}$
7. $5 \times x = 4 \times 25$
   $5x = 100$
   $\frac{5x}{5} = \frac{100}{5}$
   $x = 20$
   $\frac{5}{25} = \frac{4}{20}$
8. $3 \times y = 8 \times 21$
   $3y = 168$
   $\frac{3y}{3} = \frac{168}{3}$
   $y = 56$
   $\frac{3}{8} = \frac{21}{56}$
9. $4 \times y = 11 \times 20$
   $4y = 220$
   $\frac{4y}{4} = \frac{220}{4}$
   $y = 55$
   $\frac{55}{20} = \frac{11}{4}$
10. $20 \times n = \$120 \times 15$
    $20n = \$1,800$
    $\frac{20n}{20} = \frac{\$1,800}{20}$
    $n = \$90$
11. $\frac{15}{10} = \frac{57}{n}$
    $15 \times n = 10 \times 57$
    $15n = 570$
    $\frac{15n}{15} = \frac{570}{15}$
    $n = 38$ rolls of black and white
12. $\frac{35}{25} = \frac{n}{10}$
    $25 \times n = 10 \times 35$
    $25n = 350$
    $\frac{25n}{25} = \frac{350}{25}$
    $n = 14$ laps
13. $\frac{140}{4} = \frac{n}{6}$
    $4 \times n = 140 \times 6$
    $4n = 840$
    $n = 210$ minutes

14. $\dfrac{80}{5} = \dfrac{n}{20}$
    $5 \times n = 80 \times 20$
    $5n = 1{,}600$
    $\dfrac{5n}{5} = \dfrac{1{,}600}{5}$
    $n = 320$ points

15. $\dfrac{324}{60} = \dfrac{n}{20}$
    $60 \times n = 20 \times 324$
    $60n = 6{,}480$
    $\dfrac{60n}{60} = \dfrac{6{,}480}{60}$
    $n = \$108$

LESSON 6

PAGE 23

1. $\dfrac{3}{2} = \dfrac{n}{6}$
   $3 \times 6 = 2 \times n$
   $18 = 2n$
   $\dfrac{18}{2} = \dfrac{2n}{2}$
   $n = 9$ hours

2. $\dfrac{42}{3} = \dfrac{147}{n}$
   $42 \times n = 3 \times 147$
   $42n = 441$
   $\dfrac{42n}{42} = \dfrac{441}{42}$
   $n = 10.5$ hours

3. $\dfrac{20}{5} = \dfrac{n}{30}$
   $5 \times n = 20 \times 30$
   $5n = 600$
   $\dfrac{5n}{5} = \dfrac{600}{5}$
   $n = 120$ minutes

4. $\dfrac{7}{14} = \dfrac{n}{8}$
   $14 \times n = 7 \times 8$
   $14n = 56$
   $\dfrac{14n}{14} = \dfrac{56}{14}$
   $n = 4$ goals

5. $\dfrac{18}{1} = \dfrac{n}{9}$
   $1 \times n = 18 \times 9$
   $n = 162$ meetings

6. $\dfrac{40}{5} = \dfrac{n}{40}$
   $5 \times n = 40 \times 40$
   $5n = 1{,}600$
   $\dfrac{5n}{5} = \dfrac{1{,}600}{5}$
   $n = 320$ people

7. $\dfrac{171}{3} = \dfrac{n}{1}$
   $3 \times n = 171 \times 1$
   $3n = 171$
   $\dfrac{3n}{3} = \dfrac{171}{3}$
   $n = 57$ words

8. $\dfrac{2}{7} = \dfrac{30}{n}$
   $2 \times n = 7 \times 30$
   $2n = 210$
   $\dfrac{2n}{2} = \dfrac{210}{2}$
   $n = 105$ students

9. $\dfrac{\$390}{30 \text{ hours}} = \dfrac{n}{1 \text{ hour}}$
   $30 \times n = 390 \times 1$
   $30n = 390$
   $\dfrac{30n}{30} = \dfrac{390}{30}$
   $n = \$13$ per hour

10. $\dfrac{360 \text{ calories}}{6 \text{ servings}} = \dfrac{n}{1 \text{ serving}}$
    $6 \times n = 1 \times 360$
    $6n = 360$
    $\dfrac{6n}{6} = \dfrac{360}{6}$
    $n = 60$ calories per serving

11. $\dfrac{8 \text{ laps}}{20 \text{ minutes}} = \dfrac{n}{60 \text{ minutes}}$
    $20 \times n = 8 \times 60$
    $20n = 480$
    $\dfrac{20n}{20} = \dfrac{480}{20}$
    $n = 24$ laps

12. $\dfrac{8 \text{ calls}}{12 \text{ minutes}} = \dfrac{n}{30 \text{ minutes}}$
    $12 \times n = 8 \times 30$
    $12n = 240$
    $\dfrac{12n}{12} = \dfrac{240}{12}$
    $n = 20$ calls

13. $\dfrac{36 \text{ problems}}{60 \text{ minutes}} = \dfrac{n}{30 \text{ minutes}}$
    $60 \times n = 36 \times 30$
    $60n = 1{,}080$
    $\dfrac{60n}{60} = \dfrac{1{,}080}{60}$
    $n = 18$ problems

14. $\dfrac{12 \text{ servings}}{2 \text{ cups carrots}} = \dfrac{n}{0.5 \text{ cup carrots}}$
    $2 \times n = 12 \times 0.5$
    $2n = 6$
    $\dfrac{2n}{2} = \dfrac{6}{2}$
    $n = 3$ servings

LESSON 7

PAGE 25

1. $\dfrac{3 \text{ onions}}{1 \text{ pound}} = \dfrac{12 \text{ onions}}{n}$
   $3 \times n = 12 \times 1$
   $3n = 12$
   $\dfrac{3n}{3} = \dfrac{12}{3}$
   $n = 4$
   Twelve onions weigh 4 pounds.

2. $\dfrac{\$6}{5 \text{ fish}} = \dfrac{n}{10 \text{ fish}}$
   $5 \times n = 10 \times 6$
   $5n = 60$
   $\dfrac{5n}{5} = \dfrac{60}{5}$
   $n = 12$
   Ten fish cost $12.

3. $\dfrac{\$4.80}{12 \text{ pens}} = \dfrac{n}{10 \text{ pens}}$

   $12 \times n = 4.80 \times 10$

   $12n = 48.0$

   $\dfrac{12n}{12} = \dfrac{48}{12}$

   $n = 4$

   Ten pens cost $4.00.

4. $\dfrac{\$10}{6 \text{ roses}} = \dfrac{\$15}{n}$

   $10 \times n = 6 \times 15$

   $10n = 90$

   $\dfrac{10n}{10} = \dfrac{90}{10}$

   $n = 9$

   Nine roses cost $15.

5. $\dfrac{35 \text{ miles}}{2 \text{ hours}} = \dfrac{n}{6 \text{ hours}}$

   $2 \times n = 6 \times 35$

   $2n = 210$

   $\dfrac{2n}{2} = \dfrac{210}{2}$

   $n = 105$

   You can bike 105 miles in 6 hours.

6. $\dfrac{8 \text{ people}}{0.75 \text{ cup}} = \dfrac{12 \text{ people}}{n}$

   $8 \times n = 0.75 \times 12$

   $8n = 9$

   $\dfrac{8n}{8} = \dfrac{9}{8}$

   $n = \dfrac{9}{8} = 1\dfrac{1}{8}$

   You will need $1\dfrac{1}{8}$ cups water.

7. $\dfrac{24 \text{ calls}}{2 \text{ hours}} = \dfrac{n}{0.5 \text{ hours}}$

   $2 \times n = 24 \times 0.5$

   $2n = 12$

   $\dfrac{2n}{2} = \dfrac{12}{2}$

   $n = 6$

   You can make 6 calls.

8. $\dfrac{12 \text{ bananas}}{3 \text{ pounds}} = \dfrac{2 \text{ bananas}}{n}$

   $12 \times n = 3 \times 2$

   $12n = 6$

   $\dfrac{12n}{12} = \dfrac{6}{12}$

   $n = 0.5$

   Two bananas weigh half a pound.

9. $\dfrac{10 \text{ songs}}{20 \text{ minutes}} = \dfrac{n}{5 \text{ minutes}}$

   $20 \times n = 10 \times 5$

   $20n = 50$

   $\dfrac{20n}{20} = \dfrac{50}{20}$

   $n = 2.5$

   They play 2.5 songs.

10. $\dfrac{4 \text{ black}}{20 \text{ marbles}} = \dfrac{n}{200 \text{ marbles}}$

    $20 \times n = 4 \times 200$

    $20n = 800$

    $\dfrac{20n}{20} = \dfrac{800}{20}$

    $n = 40$

    Forty of the marbles are black.

UNIT 1 REVIEW

PAGE 27

1. $\dfrac{2}{3}$, 2 : 3, 2 to 3

2. $\dfrac{3}{7}$, 3 : 7, 3 to 7

3. 2 and 8 are even digits. 3, 5, and 7 are odd.
   $\dfrac{2}{3}$, 2 : 3, or 2 to 3

4. There are 3 vowels and 6 consonants.
   $\dfrac{3}{6}$ simplified is $\dfrac{1}{2}$, 1 : 2, or 1 to 2

5. $\dfrac{8 \text{ servings}}{56 \text{ grams fat}} = \dfrac{1 \text{ serving}}{n}$

   $8 \times n = 1 \times 56$

   $8n = 56$

   $\dfrac{8n}{8} = \dfrac{56}{8}$

   $n = 7$

   There are 7 grams of fat per serving.

6. $\dfrac{45 \text{ pages}}{30 \text{ minutes}} = \dfrac{n}{1 \text{ minute}}$

   $30 \times n = 1 \times 45$

   $30n = 45$

   $\dfrac{30n}{30} = \dfrac{45}{30}$

   $n = 1.5$

   He can read 1.5 pages per minute. He can read $1.5 \times 60 = 90$ pages per hour.

7. $3 \times 32 = 96$; $9 \times 8 = 72$
   Because $96 \neq 72$, the ratios are not proportional.

8. $12 \times 5 = 60$; $4 \times 15 = 60$
   Because $60 = 60$, the ratios are proportional.

9. $\dfrac{\$51}{3 \text{ hours}} = \dfrac{\$51 \div 3}{3 \text{ hours} \div 3} = \dfrac{\$17}{\text{hour}}$

   $\dfrac{\$36}{2 \text{ hours}} = \dfrac{\$36 \div 2}{2 \text{ hours} \div 2} = \dfrac{\$18}{\text{hour}}$

   No; Adam's rate is higher.

10. $\dfrac{12 \text{ push-ups}}{30 \text{ attempts}} = \dfrac{12 \text{ push-ups} \div 30}{30 \text{ attempts} \div 30} = \dfrac{0.4 \text{ push-ups}}{\text{attempt}}$

    $\dfrac{20 \text{ push-ups}}{50 \text{ attempts}} = \dfrac{20 \text{ push-ups} \div 50}{50 \text{ attempts} \div 50} = \dfrac{0.4 \text{ push-ups}}{\text{attempt}}$

    The rates were the same.

11. $5 \times n = 15 \times 24$

    $5n = 360$

    $\dfrac{5n}{5} = \dfrac{360}{5}$

    $n = 72$

12. $12 \times n = 10 \times 18$
$12n = 180$
$\frac{12n}{12} = \frac{180}{12}$
$n = 15$

13. $6 \times n = 8 \times 3$
$6n = 24$
$\frac{6n}{6} = \frac{24}{6}$
$n = 4$

14. $\frac{5 \text{ games}}{\$7.50} = \frac{n}{\$30}$
$7.5 \times n = 5 \times 30$
$7.5n = 150$
$\frac{7.5n}{7.5} = \frac{150}{7.5}$
$n = 20$ games

15. $\frac{24 \text{ balls}}{8 \text{ packages}} = \frac{n}{5 \text{ packages}}$
$8 \times n = 5 \times 24$
$8n = 120$
$\frac{8n}{8} = \frac{120}{8}$
$n = 15$ balls

LESSON 8

PAGE 31

1. $\frac{44}{100} = 44\%$
2. $\frac{58}{100} = 58\%$
3. $\frac{91}{100} = 91\%$
4. $\frac{30}{100} = 30\%$
5. $\frac{75}{100} = 75\%$
6. $\frac{85}{100} = 85\%$
7. $\frac{66}{100} = 66\%$
8. $\frac{92}{100} = 92\%$
9. $\frac{53}{100} = 53\%$
10. $\frac{14}{100} = 14\%$
11. $\frac{50}{100} = 50\%$
12. $\frac{17}{100} = 17\%$
13. $\frac{23}{100} = 23\%$
14. $\frac{46}{100} = 46\%$

LESSON 9

PAGE 33

1. $\frac{4}{5} = 4 \div 5 = 0.80 = 80\%$
2. $\frac{9}{10} = 9 \div 10 = 0.90 = 90\%$
3. $\frac{11}{20} = 11 \div 20 = 0.55 = 55\%$
4. $\frac{3}{50} = 3 \div 50 = 0.06 = 6\%$
5. $\frac{3}{8} = 3 \div 8 = 0.375 = 37.5\%$
6. $\frac{22}{100} = 22\%$
7. $\frac{64}{200} = 64 \div 200 = 0.32 = 32\%$
8. $\frac{14}{20} = 14 \div 20 = 0.70 = 70\%$

9. $\frac{80}{400} = 80 \div 400 = 0.20 = 20\%$
10. $\frac{17}{20} = 17 \div 20 = 0.85 = 85\%$
11. $\frac{40}{50} = 40 \div 50 = 0.80 = 80\%$
12. $\frac{28}{40} = 28 \div 40 = 0.70 = 70\%$
13. $\frac{7}{8} = 7 \div 8 = 0.875 = 87.5\%$
14. $\frac{24}{400} = 24 \div 400 = 0.06 = 6\%$
15. $\frac{16}{80} = 16 \div 80 = 0.20 = 20\%$

LESSON 10

PAGE 35

1. $55\% = \frac{55}{100} = \frac{55 \div 5}{100 \div 5} = \frac{11}{20}$
2. $95\% = \frac{95}{100} = \frac{95 \div 5}{100 \div 5} = \frac{19}{20}$
3. $15\% = \frac{15}{100} = \frac{15 \div 5}{100 \div 5} = \frac{3}{20}$
4. $44\% = \frac{44}{100} = \frac{44 \div 4}{100 \div 4} = \frac{11}{25}$
5. $83\% = \frac{83}{100}$
6. $28\% = \frac{28}{100} = \frac{28 \div 4}{100 \div 4} = \frac{7}{25}$
7. $8\% = \frac{8}{100} = \frac{8 \div 4}{100 \div 4} = \frac{2}{25}$
8. $34\% = \frac{34}{100} = \frac{34 \div 2}{100 \div 2} = \frac{17}{50}$
9. $68\% = \frac{68}{100} = \frac{68 \div 4}{100 \div 4} = \frac{17}{25}$
10. $45\% = \frac{45}{100} = \frac{45 \div 5}{100 \div 5} = \frac{9}{20}$
11. $43\% = \frac{43}{100}$
12. $15\% = \frac{15}{100} = \frac{15 \div 5}{100 \div 5} = \frac{3}{20}$
13. $68\% = \frac{68}{100} = \frac{68 \div 4}{100 \div 4} = \frac{17}{25}$
14. $29\% = \frac{29}{100}$
15. $28\% = \frac{28}{100} = \frac{28 \div 4}{100 \div 4} = \frac{7}{25}$

LESSON 11

PAGE 39

1. $0.06 = 6\%$
2. $0.64 = 64\%$
3. $0.052 = 5.2\%$
4. $0.888 = 88.8\%$
5. $0.7 = 70\%$
6. $0.005 = 0.5\%$
7. $0.0024 = 0.24\%$
8. $0.908 = 90.8\%$
9. $1.4 = 140\%$
10. $1.063 = 106.3\%$
11. $0.45 = 45\%$ students
12. $0.95 = 95\%$ students
13. $0.35 = 35\%$ drama club
14. $0.05 = 5\%$ seniors
15. $0.652 = 65.2\%$
$100.0\% - 65.2\% = 34.8\%$ people
16. $0.98 = 98\%$ students

LESSON 12

PAGE 41

1. 85% = 0.85
2. 65% = 0.65
3. 12% = 0.12
4. 48% = 0.48
5. 73% = 0.73
6. 29% = 0.29
7. 4% = 0.04
8. 34.7% = 0.347
9. 150% = 1.50
10. 2.5% = 0.025
11. 3.1% = 0.031
12. 0.3% = 0.003
13. 74% = 0.74 teens
14. 40% = 0.40 moviegoers
15. 48% = 0.48 teens
16. 5% = 0.05 teens
17. 100.0% − 62.5% = 37.5%
    37.5% = 0.375 movies
18. 120% = 1.20 effort; Accept any reasonable answer.

LESSON 13

PAGE 45

1. 82% × 50 = .82 × 50 = 41
2. 31% × 66 = 0.31 × 66 = 20.46
3. 10% × 45 = 0.10 × 45 = 4.5
4. 50% × 600 = 0.50 × 600 = 300
5. 23% × 90 = 0.23 × 90 = 20.7
6. 2.5% × 22 = 0.025 × 22 = 0.55
7. 60% × 60 = 0.60 × 60 = 3.6
8. 64.7% × 12 = 0.647 × 12 = 7.764
9. 150% × 4 = 1.50 × 4 = 6
10. 28% × 150 = 0.28 × 150 = 42 seniors
11. 75% × 32 = 0.75 × 32 = 24 actors
12. 62.5% × 40 = 0.625 × 40 = 25 stores
13. 37.5% × 24 = 0.375 × 24 = 9 fly balls
14. 70% × 140 = 0.70 × 140 = 98 houses
15. 80% × $250,000 = 0.80 × $250,000 = $200,000

LESSON 14

PAGE 47

1. 42 ÷ 50% = 42 ÷ 0.50 = 84
2. 48 ÷ 40% = 48 ÷ 0.40 = 120
3. 90 ÷ 30% = 90 ÷ 0.30 = 300
4. 15 ÷ 125% = 15 ÷ 1.25 = 12
5. 25% × n = 20
   n = 20 ÷ 25% = 20 ÷ 0.25 = 80
6. 120% × n = 18
   n = 18 ÷ 120% = 18 ÷ 1.20 = 15
7. 125% × n = 120
   n = 120 ÷ 125% = 120 ÷ 1.25 = 96
8. 25% × n = 72
   n = 72 ÷ 25% = 72 ÷ 0.25 = 288
9. 24 ÷ 30% = 24 ÷ 0.30 = 80 students applied
10. 32 ÷ 16% = 32 ÷ 0.16 = 200 businesses
11. 300 ÷ 6% = 300 ÷ 0.06 = 5,000 applied
12. 330 ÷ 165% = 330 ÷ 1.65 = 200 deer
13. 45 ÷ 37.5% = 45 ÷ 0.375 = 120 parents
14. 75 ÷ 12.5% = 75 ÷ 0.125 = 600 students

LESSON 15

PAGE 49

1. (60 ÷ 240) × 100% = 0.25 × 100% = 25%
2. (90 ÷ 240) × 100% = 0.375 × 100% = 37.5%
3. (72 ÷ 288) × 100% = 0.25 × 100% = 25%
4. (140 ÷ 200) × 100% = 0.70 × 100% = 70%
5. (18 ÷ 360) × 100% = 0.05 × 100% = 5%
6. (13 ÷ 80) × 100% = 0.1625 × 100% = 16.25%
7. (27 ÷ 50) × 100% = 0.54 × 100% = 54%
8. (6 ÷ 40) × 100% = 0.15 × 100% = 15%
9. (40 ÷ 25) × 100% = 1.6 × 100% = 160%
10. (12 ÷ 75) × 100% = 0.16 × 100% = 16%
11. (400 ÷ 1,200) × 100% = 0.333 × 100% = 33.3%
12. (36 ÷ 48) × 100% = 0.75 × 100% = 75%
13. (144 ÷ 32) × 100% = 4.5 × 100% = 450%
14. (56 ÷ 224) × 100% = 0.25 × 100% = 25%

LESSON 16

PAGE 53

1. n × 60 = 24
2. n = 0.32 × 80
3. n × 64 = 12
4. 0.80 × n = 16
5. n × 240 = 180
6. n = 0.28 × 54
7. 0.62 × n = 16
8. n × 20 = 60
9. n = 0.02 × 10
10. 0.10 × n = 12
11. n × 3.6 = 1.8
12. n = 0.022 × 200
13. 0.44 × n = 22
14. n × 68 = 16
15. n = 0.0232 × 50
16. 0.75 × n = 33

LESSON 17

PAGE 55

1. 44 is 20% of what number?
   $44 = 0.2 \times n$; $n = 220$ houses
2. 30% of 1,340 is what number?
   $0.30 \times 1,340 = n$; $n = \$402$
3. What percent of 50 (28 + 22) is 22?
   $n \times 50 = 22$; $n = 44\%$
4. 55 is 11% of what number?
   $0.11 \times n = 55$; $n = 500$ postcards
5. 85% of 8 is what number?
   $0.85 \times 8 = n$; $n = 6.8$ tons
6. What percent of 50 is 125 (175 − 50)?
   $n \times 50 = 125$; $n = 2.5\%$
7. 0.2% of what number is 18?
   $0.002 \times n = 18$; $n = 9,000$ students
8. What number is 28% of 145?
   $n = 0.28 \times 145$; $n = 40$ players
9. What percent of 623 is 386?
   $n \times 623 = 386$; $n = 62\%$
10. 94.5% of what number is 821?
    $0.945 \times n = 821$; $n = 868.8 = 868$ people

UNIT 2 REVIEW

PAGE 57

1. $\frac{47}{100} = 47\%$
2. $\frac{88}{100} = 88\%$
3. $(13 \div 20) \times 100\% = 0.65 \times 100\% = 65\%$
4. $(280 \div 400) \times 100\% = 0.70 \times 100\% = 70\%$
5. $36\% = \frac{36}{100} = \frac{36 \div 4}{100 \div 4} = \frac{9}{25}$
6. $0.42 = 42\%$
7. $61\% = 0.61$
8. $4.8\% = 0.048$
9. $42\% \times 400 = 0.42 \times 400 = 168$
10. $(40 \div 25) \times 100\% = 1.6 \times 100\% = 160\%$
11. $18 \div 40\% = 18 \div 0.40 = 45$
12. $(20 \div 20) \times 100\% = 1.00 \times 100\% = 100\%$
13. $40\% \times 80 = 0.40 \times 80 = 32$ CDs
14. $162 \div 75\% = 162 \div 0.75 = \$216$
15. $n \times 18 = 8$; $(8 \div 18) \times 100\% = 0.444 \times 100\% = 44.4\%$
16. $0.20 \times n = 6$; $6 \div 20\% = 6 \div 0.20 = 30$ hours

LESSON 18

PAGE 61

1. $60 − 57 = 3$ fewer teachers
   $3 = n \times 60$
   $\frac{3}{60} = \frac{60n}{60}$
   $n = \frac{3}{60} = \frac{1}{20} = 0.05 = 5\%$
   The percent decrease is 5%.
2. $\$60 − \$45 = \$15$ decrease
   $15 = n \times 60$
   $\frac{15}{60} = \frac{60n}{60}$
   $n = \frac{15}{60} = \frac{1}{4} = 0.25 = 25\%$
   The percent decrease is 25%.
3. $850 − 800 = 50$ seats increase
   $50 = n \times 800$
   $\frac{50}{800} = \frac{800n}{800}$
   $n = \frac{50}{800} = \frac{1}{16} = 0.0625 = 6.25\%$
   The percent increase is 6.25%.
4. $32 − 16 = 16$ acres decrease
   $16 = n \times 32$
   $\frac{16}{32} = \frac{32n}{32}$
   $n = \frac{16}{32} = \frac{1}{2} = 0.50 = 50\%$
   The percent decrease is 50%.
5. $48 − 40 = 8$ larger
   $8 = n \times 40$
   $\frac{8}{40} = \frac{40n}{40}$
   $n = \frac{8}{40} = \frac{1}{5} = 0.20 = 20\%$
   The percent increase is 20%.
6. $50 − 42 = 8$ fewer points
   $8 = n \times 50$
   $\frac{8}{50} = \frac{50n}{50}$
   $n = \frac{8}{50} = \frac{4}{25} = 0.16 = 16\%$
   The percent decrease is 16%.
7. $\$120 − \$100 = \$20$
   $20 = n \times 120$
   $\frac{20}{120} = \frac{120n}{120}$
   $n = \frac{20}{120} = \frac{1}{6} = 0.167 = 16.7\%$
   The percent decrease is 16.7%.
8. $\$85.00 − \$76.50 = \$8.50$
   $8.50 = n \times 85.00$
   $\frac{8.50}{85.00} = \frac{85.00n}{85.00}$
   $n = \frac{8.50}{85.00} = \frac{1}{10} = 0.10 = 10\%$
   The percent decrease is 10%.

9. $50 - 45 = 5$ hours less
$$5 = n \times 50$$
$$\frac{5}{50} = \frac{50n}{50}$$
$$n = \frac{5}{50} = \frac{1}{10} = 0.10 = 10\%$$
The percent decrease is 10%.

10. $40 - 32 = 8$ more workers
$$8 = n \times 32$$
$$\frac{8}{32} = \frac{32n}{32}$$
$$n = \frac{8}{32} = \frac{1}{4} = 0.25 = 25\%$$
The percent increase is 25%.

11. $\$550 - \$500 = \$50$
$$50 = n \times 500$$
$$\frac{50}{500} = \frac{500n}{500}$$
$$n = \frac{50}{500} = \frac{1}{10} = 0.10 = 10\%$$
The percent increase is 10%.

12. $300 - 220 = 80$ fewer people
$$80 = n \times 300$$
$$\frac{80}{300} = \frac{300n}{300}$$
$$n = \frac{80}{300} = \frac{4}{15} = 0.267 = 26.7\%$$
The percent decrease is 26.7%.

13. $\$440 - \$396 = \$44$
$$44 = n \times 440$$
$$\frac{44}{440} = \frac{440n}{440}$$
$$n = \frac{44}{440} = \frac{1}{10} = 0.10 = 10\%$$
The percent decrease is 10%.

14. $0.240 - 0.192 = 0.048$
$$0.048 = n \times 0.240$$
$$\frac{0.048}{0.240} = \frac{0.240n}{0.240}$$
$$n = \frac{0.048}{0.240} = \frac{1}{5} = 0.20 = 20\%$$
The percent decrease is 20%.

LESSON 19

PAGE 63

1. $\$285 - \$240 = \$45$
$$\frac{45}{285} = 0.158 = 15.8\%$$
The percent decrease is 15.8%.

2. $\$600 - \$480 = \$120$
$$120 = n \times 600$$
$$\frac{120}{600} = \frac{600n}{600}$$
$$n = \frac{120}{600} = \frac{1}{5} = 0.20 = 20\%$$
The percent decrease is 20%.

3. $\$84.00 - \$75.60 = \$8.40$
$$8.40 = n \times 84.00$$
$$\frac{8.40}{84.00} = \frac{84.00n}{84.00}$$
$$n = \frac{8.40}{84.00} = \frac{1}{10} = 0.10 = 10\%$$
The percent decrease is 10%.

4. $\$75 - \$60 = \$15$
$$15 = n \times 60$$
$$\frac{15}{60} = \frac{60n}{60}$$
$$n = \frac{15}{60} = \frac{1}{4} = 0.25 = 25\%$$
The percent increase is 25%.

5. $\$26 - \$20 = \$6$
$$6 = n \times 20$$
$$\frac{6}{20} = \frac{20n}{20}$$
$$n = \frac{6}{20} = \frac{3}{10} = 0.30 = 30\%$$
The percent increase is 30%.

6. $\$300 - \$210 = \$90$
$$90 = n \times 300$$
$$\frac{90}{300} = \frac{300n}{300}$$
$$n = \frac{90}{300} = \frac{3}{10} = 0.30 = 30\%$$
The percent decrease is 30%.

7. $\$45 - \$25 = \$20$
$$20 = n \times 25$$
$$\frac{20}{25} = \frac{25n}{25}$$
$$n = \frac{20}{25} = \frac{4}{5} = 0.80 = 80\%$$
The percent increase is 80%.

8. $\$28 - \$24 = \$4$
$$4 = n \times 24$$
$$\frac{4}{24} = \frac{24n}{24}$$
$$n = \frac{4}{24} = \frac{1}{6} = 0.167 = 16.7\%$$
The percent increase is 16.7%.

9. $\$75 - \$60 = \$15$
$$15 = n \times 75$$
$$\frac{15}{75} = \frac{75n}{75}$$
$$n = \frac{15}{75} = \frac{1}{5} = 0.20 = 20\%$$
The percent discount is 20%.

10. $\$80 - \$65 = \$15$
$$15 = n \times 65$$
$$\frac{15}{65} = \frac{65n}{65}$$
$$n = \frac{15}{65} = \frac{1}{5} = 0.20 = 20\%$$
The percent markup is 20%.

11. $\$160 - \$128 = \$32$
$$32 = n \times 160$$
$$\frac{32}{160} = \frac{160n}{160}$$
$$n = \frac{32}{160} = \frac{1}{5} = 0.20 = 20\%$$

The percent discount is 20%.
12. $120 = n \times 300$
$$\frac{120}{300} = \frac{300n}{300}$$
$n = \frac{120}{300} = \frac{2}{5} = 0.40 = 40\%$
The percent markup is 40%.
13. $\$480 - \$410 = \$70$
$70 = n \times 480$
$$\frac{70}{480} = \frac{480n}{480}$$
$n = \frac{70}{480} = \frac{7}{48} = 0.146 = 14.6\%$
The percent discount is 14.6%.
14. $\$360 - \$280 = \$80$
$80 = n \times 360$
$$\frac{80}{360} = \frac{360n}{360}$$
$\frac{80}{360} = \frac{2}{9} = 0.222 = 22.2\%$
The percent discount is 22.2%

LESSON 20

PAGE 67

1. $I = \$800 \times 3\% \times 2 = \$800 \times 0.03 \times 2 = \$48$
2. $I = \$4,000 \times 4\% \times 1 = \$4,000 \times 0.04 \times 1 = \$160$
3. $I = \$500 \times 5\% \times 2 = \$500 \times 0.05 \times 2 = \$50$
4. $I = \$700 \times 3.5\% \times 0.5 = \$700 \times 0.035 \times 0.5 = \$12.25$
5. $I = \$1,000 \times 8\% \times 2 = \$1,000 \times 0.08 \times 2 = \$160$
$T = \$1,000 + \$160 = \$1,160$
6. $I = \$900 \times 7.5\% \times 5 = \$900 \times 0.075 \times 5 = \$337.50$
$T = \$900 + \$337.50 = \$1,237.50$
7. $I = \$1,300 \times 5.4\% \times 1 = \$1,300 \times 0.054 \times 1 = \$70.20$
$T = \$1,300 + \$70.20 = \$1,370.20$
8. $I = \$800 \times 9\% \times \frac{9}{12} = \$800 \times 0.09 \times 0.75 = \$54$
$T = \$800 + \$54 = \$854$
9. $I = \$320 \times 0.06 \times 3 = \$57.60$
$T = \$320 + \$57.60 = \$377.60$
10. $I = \$900 \times 0.05 \times 0.5 = \$22.50$
11. $I = \$1,100 \times 0.09 \times 0.75 = \$74.25$
$T = \$1,100 + \$74.25 = \$1,174.25$
12. $I = \$240 \times 0.06 \times 1.5 = \$21.60$
13. $I = \$450 \times 0.005 \times 0.75 = \$1.69$
14. $I = \$1,000 \times 0.0675 \times 2.5 = \$168.75$
$T = \$1,000 + \$168.75 = \$1,168.75$

LESSON 21

PAGE 69

1. $B = 600(1 + 0.04)^2 = 600(1.04)^2 = 600(1.0816) = \$648.96$

2. $B = 900(1 + 0.06)^3 = 900(1.06)^3 = 900(1.1910) = \$1,071.90$
3. $B = 900(1 + 0.04)^5 = 900(1.04)^5 = 900(1.2167) = \$1,095.03$
4. $B = 1,800(1 + 0.06)^4 = 1,800(1.06)^4 = 1,800(1.2625) = \$2,272.50$
5. $B = 2,000(1 + 0.045)^1 = 2,000(1.045) = \$2,090$
6. $B = 3,000(1 + 0.0425)^{12} = 3,000(1.0425)^{12} = 3,000(1.6478) = \$4,943.40$
7. $B = 5,000(1 + 0.04)^5 = 5,000(1.04)^5 = 5,000(1.2167) = \$6,083.50$
8. $B = 4,250(1 + 0.06)^4 = 4,250(1.06)^4 = 4,250(1.2625) = \$5,365.63$
9. $B = 500(1 + 0.08)^{15} = 500(1.08)^{15} = 500(3.1722) = \$1,586.08$
10. $B = 300(1 + 0.01)^{12} = 300(1.01)^{12} = 300(1.1268) = \$338.04$
11. $B = 1,000(1 + 0.10)^{1.5} = 1,000(1.10)^{1.5} = 1,000(1.1537) = \$1,153.70$
12. $B = 9,000(1 + 0.03)^4 = 9,000(1.03)^4 = 9,000(1.1255) = \$10,129.50$
13. $B = 1,000(1 + 0.05)^{100} = 1,000(1.05)^{100} = 1,000(131.5012) = \$131,501.20$
14. $B = 150(1 + 0.005)^{12} = 150(1.005)^{12} = 150(1.0617) = \$159.26$

UNIT 3 REVIEW

PAGE 71

1. $30 - 15 = 15$
$15 = n \times 30$
$n = \frac{15}{30} = \frac{1}{2} = 0.50 = 50\%$
The percent decrease is 50%.
2. $30 - 15 = 15$
$15 = n \times 15$
$n = \frac{15}{15} = 1.00 = 100\%$
The percent increase is 100%.
3. $55 - 40 = 15$
$15 = n \times 40$
$n = \frac{15}{40} = \frac{3}{8} = 0.375 = 37.5\%$
The percent markup is 37.5%.
4. $225 - 200 = 25$
$25 = n \times 200$
$n = \frac{25}{200} = \frac{1}{8} = 0.125 = 12.5\%$
The percent markup is 37.5%.
5. $600 - 480 = 120$
$120 = n \times 600$
$n = \frac{120}{600} = \frac{1}{5} = 0.20 = 20\%$
The percent discount is 20%.
6. $240 - 204 = 36$
$36 = n \times 240$
$n = \frac{36}{240} = \frac{3}{20} = 0.15 = 15\%$
The percent discount is 15%.
7. $I = \$400 \times 0.045 \times 3 = \$54$

8. $I = \$1,200 \times 0.06 \times 0.5 = \$36$
9. $I = \$200 \times 0.06 \times 1 = \$12$
   $T = \$200 + \$12 = \$212$
10. $I = \$2,500 \times 0.045 \times 3 = \$337.50$
    $T = \$2,500 + \$337.50 = \$2,837.50$
11. $B = 200(1 + 0.06)^4 = 200(1.06)^4 = 200(1.2625) = \$252.50$
12. $B = 500(1 + 0.05)^3 = 500(1.05)^3 = 500(1.1576) = \$578.80$
13. $850 - 800 = 50$
    $50 = n \times 800$
    $n = \frac{50}{800} = \frac{1}{16} = 0.0625 = 6.25\%$
    The percent increase is 6.25%.
14. $60 - 44 = 16$
    $16 = n \times 60$
    $n = \frac{16}{60} = \frac{4}{15} = 0.267 = 26.7\%$
    The percent discount is 26.7%.
15. $I = \$500 \times 0.05 \times 3 = 75$
    $T = \$500 + \$75 = \$575$
16. $B = 600(1 + 0.03)^4 = 600(1.03)^4 = 600(1.1255) = \$675.30$

POST TEST

PAGES 72–73

1. $\frac{3}{4}$, 3 : 4, 3 to 4
2. $\frac{4}{7}$, 4 : 7, 4 to 7
3. $\frac{120 \text{ words}}{2 \text{ minutes}} = \frac{120 \text{ words} \div 2}{2 \text{ minutes} \div 2} = \frac{60 \text{ words}}{1 \text{ minute}}$
4. $\frac{60 \text{ miles}}{3 \text{ gallons}} = \frac{60 \text{ miles} \div 3}{3 \text{ gallons} \div 3} = \frac{20 \text{ miles}}{1 \text{ gallon}}$
5. $\frac{\$2.40}{6 \text{ bottles}} = \frac{\$2.40 \div 6}{6 \text{ bottles} \div 6} = \frac{\$0.40}{1 \text{ bottle}}$
   Each bottle costs 40 cents.
6. $\frac{\$1.95}{6 \text{ pounds}} = \frac{\$1.95 \div 6}{6 \text{ pounds} \div 6} = \frac{\$0.33}{1 \text{ pound}}$
   The price is 33 cents a pound.

7. $8 \times 16 = 128; 5 \times 10 = 50$
   $\frac{5}{8} \neq \frac{16}{10}$ No
8. $21 \times 4 = 84; 7 \times 12 = 84$
   $\frac{4}{7} = \frac{12}{21}$ Yes
9. $12n = 4 \times 3$
   $\frac{12n}{12} = \frac{12}{12}$
   $n = 1$
10. $16n = 48 \times 8$
    $\frac{16n}{16} = \frac{384}{16}$
    $n = 24$
11. $\frac{2}{7} = \frac{n}{224}$
    $7n = 448$
    $n = 64$ players
12. $\frac{\$336}{24 \text{ hours}} = \frac{n}{1 \text{ hour}}$
    $24n = \$336$
    $n = \$14$
13. $\frac{5}{16} = 0.3125 = 31.25\%$
14. $0.062 = 6.2\%$
15. $42\% = \frac{42}{100} = \frac{21}{50}$
16. $115\% = \frac{115}{100} = \frac{23}{20}$
17. $40\% \times 62 = 0.40 \times 62 = 24.8$
18. $(24 \div 60) \times 100\% = 0.40 \times 100\% = 40\%$
19. $1.4 \div 0.20 = 7$
20. $(17 \div 68) \times 100\% = 0.25 \times 100\% = 25\%$
21. $18 \div 0.25 = 72$ stamps
22. $\$21 \times 0.10 = \$2.10$ less
23. $I = \$1,200 \times 0.05 \times 3 = \$180$
24. $I = \$400 \times 0.045 \times 0.5 = \$9$
25. $B = 100(1 + 0.075)^2 = 100(1.075)^2 = 100(1.1556) = \$115.56$

Percent Triangle

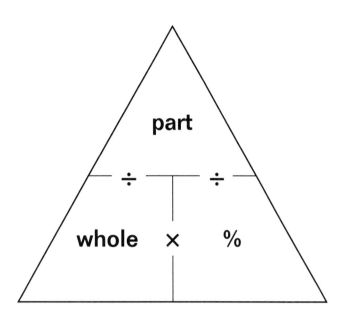

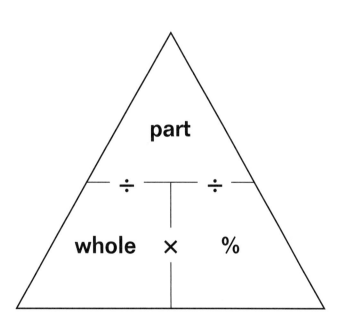

Copy Master

Percents and Problem Solving

Find the Part

    Given: the percent and the whole
    Equation: Whole × Percent = Part
    Example: Find 30% of 120.
                 120 × 0.30 = 36

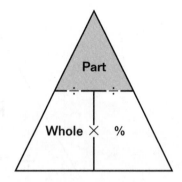

Find the Percent

    Given: the part and the whole
    Equation: (Part ÷ Whole) × 100 = Percent
    Example: 50 is what percent of 250?
                 (50 ÷ 250) × 100 = 20%

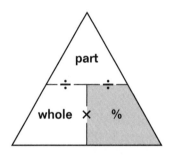

Find the Whole

    Given: the percent and the part
    Equation: Part ÷ Percent = Whole
    Example: 36 is 75% of what number?
                 36 ÷ 0.75 = 48

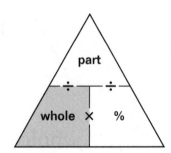

Simple and Compound Interest

Simple Interest

Simple Interest Worksheet		
Interest	I	
Principal	p	
Rate	r	
Time	t	

I	$=$	p	$\times$	r	$\times$	t

Compound Interest

Compound Interest Worksheet		
Balance	B	
Principal	p	
Rate	r	
Time	t	

B	$=$	p	$\times$	$(1$	$+$	$r)$	t

Equivalent Fractions, Decimals, and Percents

Fraction	Decimal	Percent
$\frac{1}{2}$	0.5	50%
$\frac{2}{2} = 1$	1.0	100%

Fraction	Decimal	Percent
$\frac{1}{3}$	$0.3\overline{3}$	$33.\overline{3}\%$
$\frac{2}{3}$	$0.6\overline{6}$	$66.\overline{6}\%$
$\frac{3}{3} = 1$	1.0	100%

Fraction	Decimal	Percent
$\frac{1}{4}$	0.25	25%
$\frac{2}{4} = \frac{1}{2}$	0.5	50%
$\frac{3}{4}$	0.75	75%
$\frac{4}{4} = 1$	1.0	100%

Fraction	Decimal	Percent
$\frac{1}{5}$	0.2	20%
$\frac{2}{5}$	0.4	40%
$\frac{3}{5}$	0.6	60%
$\frac{4}{5}$	0.8	80%
$\frac{5}{5} = 1$	1.0	100%

Fraction	Decimal	Percent
$\frac{1}{6}$	$0.16\overline{6}$	$16.\overline{6}\%$
$\frac{2}{6} = \frac{1}{3}$	$0.3\overline{3}$	$33.\overline{3}\%$
$\frac{3}{6} = \frac{1}{2}$	0.5	50%
$\frac{4}{6} = \frac{2}{3}$	$0.6\overline{6}$	$66.\overline{6}\%$
$\frac{5}{6}$	$0.83\overline{3}$	$83.\overline{3}\%$
$\frac{6}{6} = 1$	1.0	100%

Fraction	Decimal	Percent
$\frac{1}{8}$	0.125	12.5%
$\frac{2}{8} = \frac{1}{4}$	0.25	25%
$\frac{3}{8}$	0.375	37.5%
$\frac{4}{8} = \frac{1}{2}$	0.5	50%
$\frac{5}{8}$	0.625	62.5%
$\frac{6}{8} = \frac{3}{4}$	0.75	75%
$\frac{7}{8}$	0.875	87.5%
$\frac{8}{8} = 1$	1.0	100%

Fraction	Decimal	Percent
$\frac{1}{10}$	0.1	10%
$\frac{2}{10} = \frac{1}{5}$	0.2	20%
$\frac{3}{10}$	0.3	30%
$\frac{4}{10} = \frac{2}{5}$	0.4	40%
$\frac{5}{10} = \frac{1}{2}$	0.5	50%
$\frac{6}{10} = \frac{3}{5}$	0.6	60%
$\frac{7}{10}$	0.7	70%
$\frac{8}{10} = \frac{4}{5}$	0.8	80%
$\frac{9}{10}$	0.9	90%
$\frac{10}{10} = 1$	1.0	100%

Fraction	Decimal	Percent
$\frac{1}{100}$	0.01	1%
1	1.0	100%

Multiplying Decimals

Multiplying decimals is similar to multiplying whole numbers. However, when multiplying decimals you need to be mindful of the number of decimal places in each factor. The total number of decimal places in the numbers you are multiplying will tell you the number of decimal places in your answer.

EXAMPLE

Multiply. 28.3×6

STEP 1 Multiply.

(Ignore the decimal point for now.)

$$\begin{array}{r} {}^{4}2^{1}8.3 \\ \times 6 \\ \hline 1698 \end{array}$$

STEP 2 Count the number of decimal places in each number you are multiplying.

STEP 3 Add the number of decimal places and write the number beside the answer (product).

```
  28.3  ← 1 decimal place
×    6  ← 0 decimal places
  1698  ← 1 decimal place
```

STEP 4 Look at the number of decimal places written beside the answer. Starting at the right side of the answer, count that number of decimal places to the left, and put the decimal point there.

```
  28.3  ← 1 decimal place
×    6  ← 0 decimal places
 169.8  ← 1 decimal place
```

The product of 28.3 and 6 is 169.8.

$28.3 \times 6 = 169.8$

Dividing Decimals

Division with decimals is similar to division with whole numbers. However, you must follow some simple guidelines regarding the decimal point when dividing with decimals. If your divisor is not a decimal, but your dividend is, you simply move the decimal point in the dividend up into the quotient. If your divisor is a decimal, then you move the decimal point to the right until your divisor is a whole number. You must also move the decimal point in the dividend the same number of places.

EXAMPLE

Divide. $1.21 \div 2.75$

STEP 1 Set up the problem.

$2.75 \overline{)1.21}$

STEP 2 Count the number of decimals places in the divisor—the number you are dividing with.

The divisor has two decimal places to the right of the decimal point.

$2.75 \overline{)1.21}$
2 decimal places

STEP 3 Move the decimal point in the divisor as many places as there are decimal places in the divisor. Move the decimal point in the dividend (the number being divided) the same number of places.

Move the decimal points two places to the right.

$2.75 \overline{)1.21}$

STEP 4 Divide. Add zeroes as necessary.

```
           44
    275.)121.00
        110 0
         11 00
         11 00
             0
```

STEP 5 Look at the decimal point in the dividend. Place a decimal point directly above it in the quotient. If the decimal point falls to the left of the first digit in the quotient, place a zero to the left of the decimal point

```
          0.44
    275.)121.00
        110 0
         11 00
         11 00
             0
```

$1.21 \div 2.75 = 0.44$

RATIOS AND PERCENTS Pretest/Post Test Evaluation Chart

Name _____

Circle the number of each test item that you answered correctly.
Record how many items you got right in each section.

Pretest date _____

Post Test date _____

SECTION	PRETEST	POST TEST	LESSON REVIEW
Ratios Pretest: items 1, 2 Post Test: items 1, 2	/ 2	/ 2	Page 9
Rates Pretest: items 3, 4 Post Test: items 3, 4	/ 2	/ 2	Page 10
Learning About Unit Rates Pretest: items 5, 6 Post Test: items 5, 6	/ 2	/ 2	Page 13
Problem Solving Using Proportions Pretest: items 7, 8, 11, 12 Post Test: items 7, 8, 11, 12	/ 4	/ 4	Page 25
Converting Fractions and Percents Pretest: items 13, 14, 15, 16 Post Test: items 13, 14, 15, 16	/ 4	/ 4	Pages 33 & 35
Creating Percent Equations from Word Problems Pretest: items 17, 18, 19, 20, 21, 22 Post Test: items 17, 18, 19, 20, 21, 22	/ 6	/ 6	Pages 55
Simple Interest Pretest: items 23, 24 Post Test: items 23, 24	/ 2	/ 2	Pages 67
Compound Interest Pretest: item 25 Post Test: item 25	/ 1	/ 1	Pages 69
Total Correct	/ 25	/ 25	

Concentrate on any section in which you missed one or more items.